Is It Safe?

1/14

Is It Safe?

BPA and the Struggle to Define the Safety of Chemicals

SARAH A. VOGEL

University of California Press

BERKELEY   LOS ANGELES   LONDON

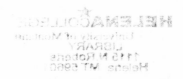

University of California Press, one of the most distinguished
university presses in the United States, enriches lives around the world
by advancing scholarship in the humanities, social sciences, and natural
sciences. Its activities are supported by the UC Press Foundation and
by philanthropic contributions from individuals and institutions. For
more information, visit www.ucpress.edu.

University of California Press
Berkeley and Los Angeles, California

University of California Press, Ltd.
London, England

Library of Congress Cataloging-in-Publication Data

Vogel, Sarah A. (Sarah Ann), 1974–
  Is it safe? : BPA and the struggle to define the safety of chemicals /
Sarah A. Vogel.
    p.   cm.
  Includes bibliographical references and index.
  ISBN 978-0-520-27357-3 (cloth : alk. paper)—ISBN
978-0-520-27358-0 (pbk. : alk. paper)
  I. Title.
  [DNLM: 1. Phenols—toxicity.   2. Economics.   3. Endocrine
Disruptors—toxicity.   4. Environmental Exposure—legislation &
jurisprudence.   5. Environmental Exposure—standards.   6. Politics.
QV 627]
  615.9′5131—dc23   2012026476

Manufactured in the United States of America

20   19   18   17   16   15   14   13
10   9   8   7   6   5   4   3   2   1

In keeping with a commitment to support environmentally responsible
and sustainable printing practices, UC Press has printed this book on
50-pound Enterprise, a 30% post-consumer-waste, recycled, deinked
fiber that is processed chlorine-free. It is acid-free and meets all
ANSI/NISO (Z 39.48) requirements.

In memory of my mother, Judy Vogel

Contents

Illustrations

Abbreviations

ACC	American Chemistry Council
AIHC	American Industrial Health Council
APA	Administrative Procedures Act (1946)
APC	American Plastics Council
BPA	bisphenol A
CDC	Centers for Disease Control and Prevention
CERHR	Center for the Evaluation of Risks to Human Reproduction
DES	diethylstilbestrol
ECHA	European Chemicals Agency
EDSTAC	Endocrine Disruptor Screening and Testing Advisory Committee
EFSA	European Food Safety Authority
EPA	U.S. Environmental Protection Agency
FDA	U.S. Food and Drug Administration
FFDCA	Federal Food, Drug and Cosmetic Act (1938, 1958)
FQPA	Food Quality Protection Act (1996)
GAO	General Accounting Office (as of 2004, Government Accountability Office)
GLP	good laboratory practices
GRAS	generally recognized as safe
HEW	Department of Health, Education and Welfare
IARC	International Agency for Research on Cancer
IBT	Industrial Bio-Test

IJC	U.S.-Canadian International Joint Commission
IPCC	Intergovernmental Panel on Climate Change
IRLG	Interagency Regulatory Liaison Group
LOAEL	lowest observed adverse effect level
MCA	Manufacturing Chemists' Association
NAS	National Academy of Sciences
NCI	National Cancer Institute
NIEHS	National Institute of Environmental Health Sciences
NIH	National Institutes of Health
NIOSH	National Institute for Occupational Safety and Health
NOAEL	no observable adverse effect level
NRC	National Research Council
NRDC	Natural Resources Defense Council
NTP	National Toxicology Program
OIRA	Office of Information and Regulatory Affairs
OMB	Office of Management and Budget
OSHA	Occupational Safety and Health Administration
OSHAct	Occupational Safety and Health Act
OTA	Congressional Office of Technology Assessment
PBDEs	polybrominated diphenyl ethers
PCBs	polychlorinated biphenyls
PFOA	perfluorooctanoic acid
PVC	polyvinyl chloride
PVP	polyvinylpyrrolidone
REACH	Registration, Evaluation, Authorisation, and Restriction of Chemicals
SDWA	Safe Drinking Water Act (1974, 1996)
SOM	Sensitivity of Method
SPI	Society of the Plastics Industry
TASSC	The Advancement of Sound Science Coalition
TSCA	Toxic Substances Control Act (1976)
WHO	World Health Organization
WWF	World Wildlife Fund, U.S.

Measurements

mg/kg	milligrams per kilogram or parts per million (ppm)
μg/kg	micrograms per kilogram (parts per billion; ppb) = 0.001 mg/kg
ng/kg	nanograms per kilogram (parts per trillion; ppt) = 0.000001 mg/kg

Preface

This book asks the question: What makes a chemical safe?

The rapid proliferation of petrochemical compounds—used to make pesticides, plastics, drugs, and many other products of modern life—over the past sixty years has transformed our global economy and ecology, as well as human understanding of our brave new world. From the mid–twentieth century onward, we marched forward in the petrochemical revolution with the assurance that the ease, health, and convenience provided by new pesticides and plastics would outweigh their risks. We were armed with comforting colloquialisms, such as "All things in moderation." A bit of chemical exposure could be tolerated, given the enormous benefits petrochemicals afforded society. And indeed, as U.S. chemical production grew, so too did the gross domestic product.

The neat simplicity and seemingly commonsense notion of this logic, however, is being contested today by emerging understandings about the risks of industrial chemicals found within the human body. Recognition that exposure is so ubiquitous that industrial compounds can now be found in umbilical cord blood and amniotic fluid alters our perspective on what defines chemical "moderation" in the twenty-first century.

In the early twentieth century, concerns about chemical hazards focused on high exposures in the workplace, and safety was defined as exposure reduction through such measures as personal protection and hygiene. Beginning in the mid-twentieth century, the burgeoning new field of environmental health research began to integrate developmental biology, genetics, toxicology, endocrinology, and biochemistry, and to explore the biological effects of the low levels of chemicals found in everyday products and the environment. Researchers now are exploring how chemicals turn genes on and off, alter development of reproductive organs, affect neurological development and behavior, and disrupt the hormonal system.

The debate about the safety of bisphenol A (BPA), a chemical primarily used in plastics production, is central to this chemical narrative. BPA's production has risen precipitously since the 1950s as the plastics market has rapidly expanded. Because of its economic success, BPA became omnipresent in the environment, and ultimately it was subjected to greater scrutiny by researchers studying the interaction of chemicals and human development and disease. Today a growing chorus of researchers is sounding the alarm about BPA because of concerns that even very low levels of exposure may increase the risk of chronic diseases including prostate and breast cancer, reproductive abnormalities, diabetes, and cardiovascular disease, as well as behavioral abnormalities. Some U.S. states have issued bans on BPA in certain products. Yet industry producers maintain that human exposure to BPA is safe, and regulators in the United States and in Europe remain largely in agreement with industry.

Is It Safe? details the political, legal, and scientific conflicts and negotiations involved in defining a chemical's safety from the 1950s to the contentious debates about BPA today. The book traces moments of crisis when changing scientific knowledge, legal contestations, and public perceptions of risks challenged markets for lucrative chemicals such as BPA. The story of BPA demonstrates how the meaning of safety is established, maintained, and subsequently challenged and contested through conflicting scientific research, values, and positions of political-economic power.

Like much of the history of chemical production and regulation, the tale of BPA reveals that chemical producers have demanded and sought consistency in the meaning of safety in order to effectively develop and maintain their markets. Their effort to maintain the stability of "safety" has collided time and time again with advancements in scientific knowledge and popular demands to protect public health.

Given that exposure to BPA and hundreds of other industrial chemicals appears to be ubiquitous, the pressing challenge for regulators, policy makers, and the public health community today is how to allocate necessary and appropriate resources to effectively manage chronic disease risks from environmental exposures to industrial chemicals. While the question of BPA's safety remains unresolved today, and there is not yet a clear path for informed decision making, this contemporary case points to the central conclusions of *Is It Safe?*

Safety is defined in a dynamic matrix of political, economic, legal, and scientific change, and, as a result, it can never be completely stable. Yet stability is exactly what chemical producers demand in order to open new markets and to keep them growing. This tension will persist, and it will continue to provide an incentive for private industry to shape the direction and interpretation of scientific research. Regulatory decision makers must navigate this unavoidable tension and, in so doing, provide consistency, clarity, and assurance that chemical exposures do not significantly undermine public health. The daunting task of balancing these competing interests—often in the context of complex scientific uncertainty, political dissonance, and conflicting pressures—demands robust processes to ensure the clarity and impartiality of the production of scientific research and the translation and interpretation of that research for decision making. If these principles of democracy are not embedded in scientific practice, defining chemical safety may protect only markets and not public health.

Time and time again, when I have shared this project with others, they asked me one simple question: "Well, what do you think? Is it safe?" That question gave rise to the title of this book, which is an effort to explain why the answer must be more than "Yes" or "No," more than the simple solution of by buying a BPA-free baby bottle. The answer requires a transformation in the way information about chemical risks is gathered and directed toward public health prevention. In a world where we have truly become what we make, the assumption that guided us into the petrochemical age—"A little bit can't hurt"—cannot guide us to a healthy and sustainable world in the twenty-first century.

Acknowledgments

This book began as my dissertation in 2006. At that time, I never could have anticipated that the story of BPA would evolve into the global debate and policy issue that it has become today. As the conflict developed over the years, the question of the book shifted: I wished to more closely consider the meaning of a chemical's safety. Everyone wants to know if BPA is safe or not, and few seem to be patient enough to consider what *safe* might mean.

It's humbling to realize how many years have passed since I began this project. In that time, I have met, interviewed, and talked and argued with many people who all, in some way, ultimately shaped this book. As with any long project, there are many people to thank. But I must begin by acknowledging a number of people without whom I know for certain that I never would have able—fiscally, mentally, or emotionally—to complete this book. For the past three years, I worked on this book while serving as a program officer for the Johnson Family Foundation. I am deeply indebted to its executive director, Andrew Lane, who supported me in this endeavor, and to its board trustees, James Johnson, Tom Johnson, Jesse Johnson, Asa Johnson, Jane Johnson, Mary Tyler Johnson, Paul Hokemeyer, and Ina Smith.

While I was studying at Columbia University, David Rosner and Elizabeth Blackmar were the best dissertation advisors a girl could ask for, and without their brilliant minds this project would have strayed far afield. I feel incredibly fortunate to have studied under them and to consider them friends. My dissertation work, and subsequently much of the research for this book, was very generously supported by the National Science Foundation, the Public Entity Risk Institute, and the American Council of Learned Societies/Andrew Mellon Foundation.

While completing this book, I was lucky enough to have George Perkovich in my life. George read every chapter more times than I can count. He gave me a depth of belief in myself and this project without which I would have abandoned it long ago. He also became my best friend and my husband.

In conducting my research, I traveled to many archives, including the papers of James Delaney at the State University of New York in Albany, the Society of the Plastics Industry papers at the beautiful Hagley Museum and Library, the papers of Wilhelm Hueper at the National Library of Medicine, the EPA Chemical Library (which no longer exists in brick and mortar), and the papers of Theo Colborn (which reside in her basement in Paonia, Colorado, and which, I hope, will one day find a proper home in an archive). I interviewed numerous researchers, former regulators, and industry representatives. I must thank them all for their patience and time spent explaining difficult concepts and theories. Some industry officials chose to speak with me off the record. Jerome Heckman, lawyer for the plastics industry for decades, generously spent several hours answering my questions and shared a white paper detailing his professional life. Theo Colborn opened her home and her personal files to me, unrestricted. She also took me on a most memorable drive through an awesome mountain pass high up in the Rockies.

As the contemporary debate about BPA unspooled and grew increasingly contentious, I frequently found myself in the middle of heated disputes. For my part, I have tried to translate the larger implications of this seemingly very narrow scientific debate for my readers, and to explain its stakes rather than argue for one side or the other. Only once, in the midst of a presentation, was I yelled at (and very rudely) by a high-level representative of a major BPA producer. My presentation focused on why emerging research presents important policy questions and challenges. The vituperative response I received only confirmed the importance of the arguments on BPA. It also helped to thicken my skin and strengthen my confidence. So I must thank that person as well, if only for his belligerence.

Over the years, I have participated in numerous conferences, meetings, and workshops about chemical policy and BPA. These experiences informed my work, my views, and, no doubt, my biases. I have listened to researchers clearly explain why they have serious concerns about human exposure to BPA due to the unsettling effects they've seen in exposed laboratory animals. I've heard other researchers express their doubts about the extent of BPA's dangers to humans. Others have expressed their deep concerns about how politically divisive the debate over BPA has become.

While some blame tougher industry tactics, others blame public interest groups. From my own experience, it's clear that the regulators stuck in the middle are not always a happy bunch and, as Daniel Carpenter notes, are often at a loss to please anyone.[1]

Moments of ugliness in this contemporary debate worry me, as they reflect the contemporary climate change debate, in which enlightened reasoning and respectable scientific arguments have been obfuscated by willful ignorance and theocracy. Yet the recommendations with which I end this book with are modest. At the broadest level, they call for efforts to sustain democracy in science by enhancing clarity, independence, and integrity in research and the translation of evidence for decision making. More specifically, they point to the need for systematic reviews of environmental health research and transparent recommendations for decision making based on that research.

Daniel Fox introduced me to the work of Tracey Woodruff and Patrice Sutton at the University of California, San Francisco, who are working to adapt systematic review methodologies to environmental health research. It has been a privilege—and great fun—getting to know them both. I am also indebted to Fox for agreeing so many years ago to read a draft of my dissertation. He worked with me to develop this book and introduced me to the University of California Press and to Hannah Love, its editor for public health, who has been a joy to work with.

Introduction

Jerome Heckman is a short, rotund man with oversized glasses. Now in his mid-eighties, Heckman still works in his large office at Keller and Heckman, the global law firm he established and expanded over the past half century, representing the world's leading producers of plastics, pesticides, food additives, and other chemical specialty items. As the general counsel for the Society of the Plastics Industry (SPI), the industry trade association, Heckman has led plastics producers and manufacturers through numerous crises that threatened to stall or close markets: from efforts to ban plastic bottles to concerns about the toxicity of vinyl chloride, styrene, and, today, bisphenol A (BPA), a chemical used in polycarbonate plastic and epoxy resins. His success and skill in navigating through these potential market barriers has earned him godlike status among his clients at the SPI.[1]

Looking back on his career, Heckman recalled a story of how the young plastics industry and his law firm overcame a crisis concerning styrene in the early 1960s. A major chocolate mint manufacturer had planned to announce at a press conference the release of its exciting new package made from the plastic polystyrene. Yet when the company president sampled a mint, he tasted not the sweetness of the chocolate mint, but the surprising and distinct flavor of gasoline. Imagine his surprise. As it turned out, styrene had migrated from the packaging into the fatty chocolate, unpleasant evidence that chemical constituents of the plastic had contaminated the food. The mints were dumped in an open lot in Long Island and were promptly eaten by neighborhood children. Subsequent reports from angry parents and newspaper stories of the incident eventually captured the attention of officials at the Food and Drug Administration (FDA), and its assistant commissioner called Heckman to inform him that the agency might have to pull the plastic's approval. For several years, Heckman

1

worked with the trade association to coordinate the industry's response, which included determining an industrywide level at which styrene would be allowed to migrate from polystyrene—a migration limit. Once the trade association had internally agreed on a migration limit that all producers could meet, Heckman informed the FDA that it could publish this limit as an acceptable and feasible safety standard.[2] Polystyrene stayed on the market and remains in production today despite strong evidence of styrene's carcinogenicity.[3]

Plastics are ubiquitous today in part because Jerome Heckman has done his job. His long career has had one major objective: to open and keep open markets for plastics.[4] A key factor in keeping markets open has been to secure the safety of plastics and the chemicals used to make them. This has not always been easy. Many chemicals used in plastics production have introduced new hazards and health risks to humans and the environment. As plastics have proliferated in the market, the chemicals used in their production have leached into food, entered the soil, air, and water, and migrated into the human body. Since the 1950s, scientists, consumers, and public health advocates have raised concerns about the risks of plastics pollution and the hazardous chemicals used in plastics production. By the 1960s and 1970s, plastics and the petrochemicals used to make them were turning up in unexpected places. Spilling into oceans in the form of pellets (a preproduction form), plastics were ending up in the bellies of sea animals. Firefighters faced toxic smoke emitted by burning plastics such as vinyl flooring and new foam insulations. The chemical constituents of plastic food wraps and containers contaminated foods.[5]

"More recently," Heckman wrote in the early 2000s, "we have been in the vanguard with the SPI Committee to try to fend off the questionable attacks made on Bisphenol A, the monomer used to make polycarbonates and many can coatings, and on various additives used to make other polymers function as intended. The science has become more sophisticated, but the goal for us has always been the same—open markets and keep them open for the plastics industry, while making sure the public health is fully protected."[6]

Opening markets and keeping them open required (and continue to require) that plastics and their chemical constituents be deemed safe. But what does *safe* mean? Who decides, and on what basis? For Heckman, the answer is simple: plastics are safe. While chemicals used to make plastics, and to provide them with flexibility and resiliency to sunlight and fire, do migrate into the environment and into humans, the plastics industry has long contended that they do so at such small amounts that they pose little health risk. In other words, styrene and vinyl chloride may be carcinogenic and BPA toxic to reproductive development at high levels, but at low levels they can be safe. This logic builds upon the founding principle of toxicology, which states, "The dose makes the poison." Given that risk is inevitable and unavoidable in life, the rapid proliferation of synthetic chemicals after the Second World War meant that some exposure became an inescapable fact of modern life. The challenge for researchers, industry, regulators, and public interest representatives over the last half century has been to determine what level of risk is acceptable and what may be deemed safe in the context of ever-expanding chemical and plastics production and emerging scientific insights into the impact of these substances on our health.

THE DISPUTED DEFINITION OF SAFETY

This book is about the complex, messy, and dynamic process of defining chemical safety. Chemical safety is a human construct. It is inherently unstable and evolving because it is produced through the interactions of science and scientists; law and lawyers, lobbyists, legislators, and courts; industry and consumer advocates; chambers of commerce; and environmentalists. All of these actors represent important perspectives and interests, although their power, resources, and evidence shift with time. In its most reduced form, the contestation is between economic interest and public health, with all of the protagonists arguing that they seek both goals, not one to the exclusion of the other. In the end, the practical questions become "How safe is safe enough?" and "How certain must we be of the risks?"

Is It Safe? narrates episodes of crisis in chemical safety during the past sixty years. It chronicles how individuals, technologies, and the altered molecular environment have changed scientific understanding of the effects of chemicals on human health. It describes legal challenges to laws defining acceptable levels of exposure and explains how the modern environmental movement emerged to change public perceptions of risk by questioning petrochemical production's impact on the well-being of humans and the larger environment. It details the influence of major industry trade associations and men such as Heckman in shaping and managing legal interpretations of safety and the scientific evidence used to support those interpretations.

In sum, the story describes how the safety of chemicals has been defined in the petrochemical age. I chose to focus on the past sixty years because they mark the convergence of the "petrochemical revolution," as described by historian Alfred Chandler, and the "risk society," as defined by German sociologist Ulrich Beck.[7] The expanding production and use of petrochemicals to make value-added plastics, pesticides, and pharmaceuticals significantly transformed political, economic, environmental, and social conditions of society in such a way that regulatory decision making became largely about the distribution of risk.

Life has always been full of risks. What differentiates modern risks, as Beck explains, is that we knowingly "create them ourselves, they are the product of human hands and minds, of the link between technical knowledge and the economic utility calculus."[8] The petrochemical revolution generated new demands and responsibilities for the U.S. government to balance economic growth with the protection of public health. The government responded to chemical risks primarily by expanding regulatory authority and research. But whether these actions have effectively managed risk and protected public health is a central tension throughout this book.

This story begins in the 1950s, when petrochemicals utilized to make new plastics, drugs, dyes, and pesticides rapidly entered commercial production and transformed the economy and environment, notably in the production of food. The first chapter details scientific and political debates in the 1950s that led to the passage of the 1958 Food, Drug and Cosmetics Act Amendment (also known as the Food Additives Act), which gave the FDA greater authority to regulate the safety of chemicals in food and included the Delaney clause, which specifically restricted chemical carcinogens. During the 1950s, lawmakers, FDA regulators, and researchers wrestled with fundamental questions that would shape chemical policy for years to come: Would chemical contaminants be avoided or managed? How would safety be defined? Who would identify risks and define safety, and on the basis of which information? And, importantly, should cancer-causing chemicals (i.e., carcinogens) be allowed in food at all?

In the 1960s and 1970s, as detailed in the second chapter, the FDA began to implement and interpret the new food law, and it confronted the question of whether the Delaney clause gave the agency broad authority to ban a carcinogen per se regardless of the level of exposure from food. With the emergence of environmental and consumer health organizations and mounting public concern about cancer, regulation of carcinogens had become a hotly contested policy issue by the late 1960s. Advocates challenged the FDA's policy for defining the safety of chemical carcinogens and forced several compounds off the market. But by the 1980s, with the rise of political conservatism and the waning influence of public interest organizations, restriction on carcinogens in food started to be seen by some players as a policy aberration, an example of excessive regulatory oversight.

The second half of the book continues chronologically but turns to focus on BPA: how its safety was defined in the 1980s, and how its safety came to be challenged beginning in the 1990s, in the context of significant developments in environmental health research. BPA is a chemical originally synthesized in 1891 that has been used in plastics production since the 1950s and is now found throughout the environment and in humans.[9] It also has long been known to have estrogenlike (i.e., estrogenic) properties. The third chapter details how BPA's toxicity and carcinogenicity were assessed by federal research institutions and private laboratories and how its regulatory safety was established by the Environmental Protection Agency (EPA) in 1988. Much of this work coincided with the development of the field of environmental health research, which focused on synthetic chemicals with estrogenlike properties, such as polychlorinated biphenyls (PCBs) and DDT. This research explores how exposure to environmental estrogens early in development, even at nontoxic levels, can lead to disease later in life, and it is the foundation of the theory of endocrine disruption. First articulated in the early 1990s, the theory holds that some chemicals in the environment can interact with the hormone or endocrine system and negatively affect organisms' development and function.

The thesis was met with considerable skepticism, which foreshadowed emerging debates about the BPA's risks. Chapter 4 describes the political and scientific struggles among regulators, environmental advocates, researchers, and industry representatives to define the meaning of endocrine disruption and the subsequent policy response to it in the 1990s. Did endocrine-disrupting chemicals present novel risks to humans, and did these risks require significant changes to regulatory safety standards and regulatory toxicology? What were the low-dose risks of endocrine-disrupting chemicals?

Regulatory decisions about endocrine disruptors stalled in the early 2000s, but research in the field expanded rapidly. Chapter 5 follows the swift proliferation of research on BPA as a potential endocrine-disrupting chemical in the first decade of the twenty-first century and the efforts of government institutions, academics, and industry representatives to evaluate and interpret the emerging evidence. At the center of the debate over BPA's threat as an endocrine disruptor were questions about which research is relevant and sound for use in policy decision making, and according to whom. How should new research inform regulatory safety standards?

Finally, chapter 6 describes political responses to conflicting interpretations of BPA's safety. In recent years, environmental advocacy groups have launched campaigns to ban BPA, retailers have moved to BPA alternatives, and industry trade associations and the FDA have continued to uphold the safety of BPA. In the struggle to define BPA's safety, the influence of power and politics on the regulatory decision-making process is seen to be deeply intertwined with science. Recognizing that politics cannot be removed from science, I point out the need to better manage political power in decision making via approaches that emphasize transparency and independence.

Throughout this book, regulatory decisions are often the vortex around which competing forces and interests involved in defining safety swirl. Regulators navigate politically treacherous waters when making decisions about chemical safety, whether regarding carcinogens in the 1970s or endocrine disruptors in the 1990s. Agency leaders must consider the scientific evidence and the law as well as the external and internal political pressures that collectively define their agency's reputation. They must often manage competing political-economic pressures and demands from industry, public interest organizations, the public, Congress, the White House, and diverse agency staff. The result, as described in the following pages, is that regulatory agencies rarely please all stakeholders. As Daniel Carpenter notes in his seminal work on the history of drug regulation at the FDA, "Except for particular periods whose exceptional nature proves the rule, national political culture in the United States has often been hostile to the idea that government agencies are to be trusted."[10]

Industry trade associations and their lawyers recognized early on the critical importance of developing long-term relationships with staff within regulatory agencies in order to manage and guide the process of defining safety. Industry then lent power to agency decision making by following the rules and processes that agencies set forth, which had often been mutually agreed upon.[11] Through this process, industry trade organizations have become an essential source of scientific and legal guidance for regulators. Indeed, Heckman has called industry trade associations "quasi-public institutions."[12] Politically appointed agency leaders come and go, but the trade associations' relationships with agency staff endure.[13]

A prevailing theme throughout the book is that the definition of safety is episodically destabilized by conflict between industry's demand to keep markets open and the evolution of scientific research. If markets are to be maintained and expanded, chemicals must be either entirely exempted from regulatory oversight or defined as safe by regulators, the courts, and the consumer marketplace. When new scientific insights challenge an established regulatory standard, the definition of legitimate science itself becomes contested. Scientists have repeatedly discovered new ways in which chemicals interact with biological processes that call into question the risks and safety of exposure. Public interest groups in the 1960s and 1970s and environmental advocates from the 1980s up to today have marshaled emerging scientific research to challenge the relationship between regulators and the industries they regulate. This is seen most vividly in the contemporary dispute over which scientific evidence should be used to determine whether BPA is safe. Indeed, advancements in environmental health science over the past several decades and arguments over its use in regulatory decision making are central to understanding how safety is defined and contested today. It is, therefore, worthwhile to preview the key aspects of changing scientific research on chemicals.

ENVIRONMENTAL HEALTH SCIENCE AND SAFETY

As Heckman rightfully noted, the contemporary debate over BPA safety reflects the increasing sophistication of scientific research on chemical risks. New research in environmental health sciences has revealed widespread human exposure to industrial chemicals and has provided new insights into how chemicals influence and alter biological development and function, particularly through gene-chemical interactions and through more recent understandings of epigenetic interactions (i.e., modifications in a cell that alter gene expression without changing DNA structure). These recent discoveries reflect a historical trend in the development of environmental health sciences.

As historian Christopher Sellers describes in *Hazards of the Job*, environmental health science emerged in the Progressive era out of the practice of industrial hygiene. Like industrial hygiene, environmental health science is defined in part by the use of chemical measurements, identification of threshold levels of toxic effects, and distinctions between normal and abnormal biological responses. Yet in the post–World War II period, as Sellers details, environmental health science shifted its gaze beyond the workplace to investigate chemical exposures in the general population. The implications of this transition were twofold: researchers examined the biological effects of much lower concentrations than those observed in the workplace, and they expanded the range of endpoints (i.e., biological effects such as death, change in organ weight, malformation, and so forth) investigated well beyond poisoning, acute toxicity, and even cancer, which had long dominated the field of industrial hygiene. Today, environmental health science increasingly explores the relationship between environmental exposure and chronic disease effects, not simply acute toxicity and cancer.[14] The development of environmental health research calls into question the prevailing logic used to define the safety of chemicals for the past half century: namely, that high-dose effects can reliably predict low-dose effects and that safety can be defined simply as the absence of toxicity or carcinogenicity.

Advances in analytical chemistry since the 1950s have further destabilized the meaning of safety by allowing detection of ever-lower concentrations of chemical compounds in foods as well as within the human body—its tissue, blood, breast milk, urine, and amniotic fluid (i.e., biosamples). Biomonitoring—testing the human body for evidence of chemical exposure—became an effective method to detect lead and PCB exposure, particularly in children, in the mid-1970s. In 2001, the Centers for Disease Control and Prevention (CDC) began to issue an annual report on biomonitoring results from a National Health and Nutrition Examination Survey, the *National Report on Human Exposure to Environmental Chemicals*. Today the CDC tests for several hundred synthetic chemicals, including BPA, present in representative samples of the American population. The numbers of compounds tested is expected to increase each year, unless major budget cuts stymie this effort.[15] Biomonitoring studies paint an unsettling picture of the world in which we all live. We have become, quite literally, what we make: that is, a little bit plastic, a bit flame retardant, a bit rocket fuel, a bit lead, and so forth. Given this understanding, the obvious question is: What are these chemicals doing to our bodies?

From the late 1950s through the 1990s, safety was measured according to a chemical's ability to elicit toxic, carcinogenic, or perhaps mutagenic effects on animals at high doses of exposure. Regulators used this information to make predictions about potential effects at the low levels that might be experienced in the general human population. For most chemicals, the presumption held that at some very slight level toxic effects become negligible. Below certain levels, exposure could be deemed safe. But by the late twentieth century and up to today, the scope and breadth of adverse effects investigated and measured in studies of industrial chemicals have expanded significantly as the doses studied have continued to drop. Today researchers from diverse disciplines study how minute levels of chemicals alter gene expression, interact with hormone receptors, and alter development of organs, tissues, the brain, and the functioning of different systems, and how the timing of the exposure, as well as the dose, influences the effect observed. Research by scientists trained in genetics, endocrinology, and developmental biology explores the effects of chemical exposures on the immune, reproductive, neurological, and metabolic systems.

Over the past decade, environmental health researchers have increasingly begun to apply the tools, technologies, and theories developed from the mapping of the human genome to study chemical-gene interactions. Whereas initially this mapping promised to yield the keys to conquering chronic diseases, the emergent understanding today suggests that most diseases result from complex interactions between genes and the environment. Simply put, researchers are studying the biological effects of everyday exposures and beginning to explore how such exposures influence and direct the disease process. In a world facing rising rates of chronic diseases, researchers are asking about the relationship between everyday chemical exposures and behavioral abnormalities, diabetes, obesity, reproductive disorders, disrupted immune function, and various cancers.

Contemporary debates over the safety of BPA epitomize the destabilizing effect of advances in scientific research. Beginning in the 1990s, researchers began to explore the estrogenic effects of exposure to BPA at levels far below the purported safety level. Researchers and environmental advocates demanded reassessments of the chemical's risks and increasingly called for its restriction from the market on the basis of evidence from low-dose animal studies. The implications for markets and consumer confidence are evident in the proliferation of "BPA-free" labels now found on food containers, water and baby bottles, and toys, and in the number of chemical bans. Heckman and the plastics industry have responded with efforts to "fend off the questionable attacks made on Bisphenol A."

OF BPA, ESTROGENS, AND ENDOCRINE DISRUPTORS

BPA and the emergent debate over its safety are a central part of both this book and the history of the meaning of chemical safety for several reasons. First, the history of BPA's production and use traces that of the petrochemical revolution and its impact on the environment and regulation. The production of BPA, an industrial chemical used primarily in plastics since the 1950s, grew precipitously as markets expanded. By the early twenty-first century, BPA's production in the United States had topped two billion pounds, and global production hovered at six billion pounds per year, making it one of the most highly produced chemicals in the world.[16]

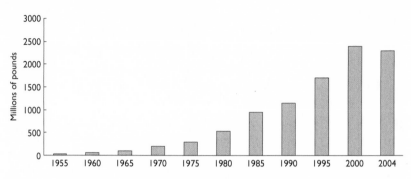

Figure 1.   Bisphenol A production in the United States.

It has been used predominantly in the production of epoxy resins, which are strong adhesives, and polycarbonates, which are hard, heat-resistant, clear plastics. BPA is also used in thermal papers (e.g., carbonless receipts and fax paper) and as a flame retardant. The commercial success of BPA meant that it found widespread use in consumer products, which in turn brought it to the attention of regulators and researchers.

Second, BPA has long been presumed to be safe in both the workplace and the U.S. food supply. The chemical was studied fairly extensively as part of the expansion of the federal government's efforts to test chemicals beginning in the mid- to late 1970s. In this respect, BPA was not an unknown chemical whose presence in the market and environment indicated a failure to gather information and set safety standards. Beginning in the 1990s, research on BPA raised new questions directly related to its estrogenic activities at low levels of exposure, which introduced new complexities regarding simple assumptions about how it functions (or does not) as an estrogen and the long-term health effects of chronic, everyday exposure in humans. In this respect, the story of BPA tells how a chemical long presumed to be safe suddenly became an "endocrine-disrupting chemical," and what this meant to regulatory policy and the conceptualization of safety.

Recognition that some industrial chemicals exhibit hormonelike properties was not new. In the 1920 and 1930s, at a time when the newly formed field of biochemistry held the promise of therapeutic "magic bullets" for disease treatment—notably the isolation and synthetic production of insulin and later estrogen—the biochemist and medical doctor Edward Charles Dodds, working at the University of London Middlesex Medical School, first identified BPA's ability to exhibit estrogenlike properties. Dodds's objective at the time was to identify a potent synthetic form of estrogen to use as a drug. In the process he identified dozens of compounds, some of which were manmade, that exhibited estrogenic properties, among them BPA and the potent synthetic estrogen diethylstilbestrol (DES).[17]

When DES came onto the market in the 1940s, estrogens already were used to treat menopausal symptoms and a wide array of purported diseases of aging, as Elizabeth Siegel Watkins describes in *The Estrogen Elixir*, a history of the use of hormone replacement therapy.[18] Watkins traces understanding of the risks and benefits of estrogen therapy and, subsequently, its rise and fall and rise again over the past sixty years. Conflicting evidence of estrogen's carcinogenicity and reproductive toxicity on the one hand, and its potential role in prevention of disease, including heart disease and cancer, on the other, has perplexed researchers, physicians, regulators, and laypeople for decades. As Watkins succinctly remarks, "Simply put, too much was expected of estrogen."[19] And now it appears too little regard was given to its complexities.

DES was a cheap drug founded on the expectation of estrogen's therapeutic miracles. Despite uncertainty about its medical value and concerns over estrogen's carcinogenicity, DES was immediately used to treat menopausal symptoms and to prevent miscarriage in pregnant women. It was also used as a feed additive for poultry and livestock to increase meat production. Then, in the early 1970s, DES was found to cause a rare vaginal cancer in the children of women who took the drug during their pregnancies. Subsequent research found reproductive abnormalities in men and women exposed to DES during fetal development.[20]

The study of the delayed, devastating effects of DES strongly influenced the study of the carcinogenicity and reproductive effects of estrogenic compounds and, later, endocrine-disrupting chemicals in the 1980s and early 1990s. This history was first told in a popular book on endocrine disruption, *Our Stolen Future*.[21] DES became a central trope in the story of endocrine disruption. The basic argument is that research on DES and other synthetic chemicals with hormonelike properties demonstrates that by interacting with hormones these chemicals can result in health effects even at very low levels of exposure if that exposure occurs during critical periods of development in an organism. With DES, exposure early in fetal development increases the risk of reproductive abnormalities and cancer. In his book *Hormonal Chaos*, Sheldon Krimsky, a philosopher of science, details the emergence of the theory of endocrine disruption from research on DES and environmental estrogens and the new challenges these findings present to understanding the risks of chemical exposures.[22]

In her recent book *Toxic Bodies*, historian Nancy Langston extrapolates on the historical relationship between DES research and the theory of endocrine disruption to draw lessons for regulatory decision making today. Specifically, she explores why the FDA failed to make a precautionary decision, despite the risks of DES recognized in the 1940s and 1950s, and allowed the drug to remain on the market until the 1970s. She highlights how political pressure, social norms about sex and gender, and economic demands allowed regulators to push aside what in retrospect were early warnings of DES's risks.[23] Langston's thesis builds on an argument articulated by Dolores Ibarreta and Shanna Swan in a 2002 series of essays, *Late Lessons from Early Warnings*, where they contend, "The DES story demonstrates that long-term and hidden effects of hormonal exposure . . . are possible, and that such consequences may be devastating. Therefore, extreme caution should be taken before exposing pregnant women to substances that may alter the endocrine system."[24] Langston draws parallels between the FDA's failure to restrict DES and current debates about endocrine disruptors. "Learning the lessons of DES," she suggests, "can help us address current disputes over regulating today's endocrine-disrupting chemicals." The resounding lesson, in this view, is that waiting for overwhelming evidence of harm delays necessary action. By taking a precautionary approach to endocrine-disruptor evidence, one might avoid unnecessary harm.[25]

Is It Safe? contributes to the literature on the history of endocrine disruption and the regulatory debates about DES and other chemicals. As described in its detailed account of the contested safety of BPA, this book argues that the multiplicity of interests and the complexity of scientific understandings involved in decision making about chemical safety do not easily allow for precautionary action. Most important, merely defining what constitutes an endocrine-disrupting chemical has stalled any regulatory action for over a decade in the United States. How does one make a precautionary decision about what to produce or restrict without a shared process to identify endocrine disruptors? For instance, if a manufacturer seeks to be cautious and remove endocrine-disrupting chemicals from its products, which information and levels of evidence should it use to determine whether a chemical is or is not an endocrine disruptor? How should that manufacturer compare evidence about endocrine disruption with that of carcinogenicity? For better and for worse, the complicated process of developing a shared definition and method to evaluate evidence is essential for making any decisions, including those that seek to best protect public health. This definition and methods are still evolving and are the subjects of competing claims about scientific legitimacy and what constitutes sound science today.

Highlighting such complexities and realities of decision making is not the same as saying that personally I don't find the hormonally active agents running through my body alarming and deeply unsettling. I have given no consent to be exposed. Nor has my friend who is treating meta-static breast cancer in her early thirties. The challenge I see facing regulators, researchers, and industry today is in determining which new tools, technologies, methods, and legal standards are needed to identify significant public health risks of chronic disease development from everyday exposures. These include statutory changes and regulatory reforms; new tools and technologies that can help better inform which chemicals will be produced and manufactured; and new evaluation methods to improve the transparency and independence of how evidence is interpreted and used to make public health decisions. Different levels of evidence—what is certain, less uncertain, and uncertain—can be used to guide decisions about what to restrict or avoid, in which populations, and, conversely, what *to* produce and manufacture. It is not enough to simply say, "No endocrine disruptors can be produced." To actually improve health and move society toward cleaner production requires a shared understanding of endocrine disruption and the necessary research questions, decision-making tools, and frameworks to adequately evaluate these risks.

The contemporary debate about BPA's safety—which involves U.S. and global policy makers, scientists, industry, and environmental advocates—embodies a number of critical issues confronting the future of chemical regulation and production. These include determining how to improve the clarity and independence of interpretation of scientific evidence and how to better manage the powerful interests involved in defining safety. The debate also raises questions about which biological effects of exposure constitute adverse events. Are the right questions being asked about the risks of human exposure to synthetic chemicals? Do standard methods for assessing risk sufficiently account for vulnerable periods of exposure and the long-term effects of chemicals at the concentration levels detected in humans?

Yet as this book repeatedly shows, because stable understandings of safety are necessary to inform future markets, shifts or revelations in scientific understandings of chemicals are often resisted and the interpretation of scientific research is deeply politicized. Therefore, if chemicals are to be made safer or cleaner, we must consider not only which information is needed to make this determination but also who conducts the research and, importantly, who interprets the results. Which research questions and data are needed, and what does this information mean? What *is* an adverse effect, for legal and regulatory purposes? What are the appropriate tests and study designs? How much evidence is necessary to make a decision? In the battle to determine BPA's safety, all these considerations for defining safety are in play.

If this book has one simple objective, it is to focus critical attention on the meaning of safety. Too frequently the word is used in chemical regulation and research, in public hearings, and in newspaper articles without careful attention to its meaning. Safety is too often viewed as an absolute, when in practice it is a dynamic and often elusive concept. Safe for whom? Safe in which context? Safe on the basis of which values and preferences? Safe according to which evidence and which uncertainties? Failure to scrutinize safety's layers of meaning can lead to a simplified notion that the concept involves only scientific evidence. If this were true—wishful thinking goes—science could be separated from politics, law, and commercial interest. This purified science would direct decision making about what is and is not safe and would end the messiness of debate and dispute. Getting the science right—making it "sound" with more and better data—would then, purportedly, end all political controversy. But if this book shows anything, it is that ensuring a more democratic process to define safety demands effective and transparent management of the politics and value preferences within scientific practice and between scientific practice and politically powerful individuals and institutions. Ensuring safer and more sustainable production in the future demands as much attention to these relationships of power and politics as to science itself.

I began this book as an exploration of how it came to be that we all contain a bit of plastic. For me, it ended with the recognition that we always will be a reflection of the world that we make. Key to building a healthy society is recognizing that chemical safety is not just a matter of scientifically determining which chemicals and materials to use and not to use, or how to use them. A healthy society demands that what we know, and how we know it, is mediated through processes that seek to uphold transparency and independence and to manage major imbalances in political and economic power, interests and values. To protect public health, the definition of safety, like that of a healthy society, must be crafted within a just political and economic system. This is not a new argument, but it has become increasingly salient as political debates on, for example, climate change move further afield of scientific evidence. Like the concept of sustainability, safety is not an absolute state. There are no magic bullets, no simple methods to build a healthy and just society. Instead, we must rely on human innovation and imagination, and on the relentless pursuit of knowledge and justice.

1 Plastic Food

Babies are not just little adults. This simple observation explains why the Beech-Nut Company, a manufacturer of baby foods, voluntarily chose to conduct safety tests of pesticide residues in its products. According to a company representative, its tests considered "prenatal, environmental, physiological, and structural [factors]—which may cause a baby to react to food residues in a manner different from the adult." As a reason for Beech-Nut's tests, the representative pointed to the pesticide DDT and evidence that its "estrogen-like" qualities negatively affected sexual development in young animals. Given the vulnerabilities of babies, the company established a near-zero tolerance for DDT in its products. These statements sound contemporary, but they were made more than fifty years ago at a congressional hearing on the risks of industrial chemicals in the food supply. The company bore the costs of and responsibilities for testing; in the early 1950s, the federal government required no premarket safety testing for the hundreds of new chemicals transforming food production in America of the sort that was required for new drugs.[1]

Beech-Nut and other U.S. private food producers were essentially managing the potential risks of human exposure to hundreds of new industrial chemicals without any roadmap or rules. This came at a high price, at least for Beech-Nut, which spent thousands of dollars developing and conducting tests. The company was in a bind. On the one hand, consumers were demanding greater assurance of the safety of baby food, while on the other hand, the company's competitors found Beech-Nut's actions "hysterical."[2] And so in the early 1950s, a Beech-Nut representative came before members of Congress to ask that the government fix this problem by setting tolerance limits or safety standards for all pesticides.[3]

By the early 1950s, innovations from the Second World War, such as DDT and new plastic products such as Saran Wrap, were rapidly entering commercial markets. The speed and scope of commercial use of pesticides and plastics, specifically those derived from an abundant and cheap supply of petroleum, brought tremendous change to the American economy and environment. The petrochemical industry—the integration of the petroleum and chemical industry—had become the critical building block of a booming consumer economy. It promised abundance, ease, and prosperity, but it also introduced unprecedented hazards into the environment and into the human body.

A decade before Rachel Carson's *Silent Spring* (1962) popularized the dangers of pesticides, a small but vocal collection of consumers, scientists, and sympathetic politicians raised serious concerns about the long-term health impacts of pesticides, plastics, dyes, and many other classes of petrochemicals (derivatives of oil created when making gasoline) increasingly making their way into the American food supply. Over the course of the 1950s, Congress debated the need and, by the end of the decade, the scope of reforms to the 1938 Federal Food, Drug and Cosmetics Act to address new chemical hazards. The questions that emerged in these legislative debates considered how the state would manage and mitigate the seemingly inevitable risks of the new petrochemical age. What defined a chemical's safety if use and exposure were integral aspects of an advanced technological society? Answering this question shaped the political and scientific contours of debates in chemical policy for the next half century.

THE PETROCHEMICAL REVOLUTION

The 1950s marked the beginning of an unprecedented economic boom in the United States. The stable provision of cheap oil provided the foundation for this economic and material transformation. Securing a cheap and reliable oil supply required several elements: exploration of and investment in domestic reserves, government pricing controls—supported by administrations from Theodore Roosevelt to Jimmy Carter—and development of stable relations with oil-producing countries. In *The Prize*, a history of oil, Daniel Yergin argues that by the 1950s a new "petroleum world order" had emerged that secured a steady supply of cheap oil to the burgeoning American consumer economy.[4] As world oil prices fell, domestic consumption grew rapidly—from 5.8 million barrels per day in 1949 to 16.4 million per day in 1972.[5]

With the end of gasoline rationing in 1945, American drivers hit the road. Americans owned twenty-six million cars by the end of the war, and only five years later that number had nearly doubled.[6] As the cold war heated up, the federal government continued to expand military contracts that had driven production during the war, thereby allowing the military to maintain its position as a major purchaser of U.S. goods and services. The expansion of the defense industry created new jobs, and government policies kept inflation and prices low, in particular the price of gasoline.[7] In 1956, Eisenhower signed the Federal Highway Act, which allocated millions of dollars to the expansion of the interstate highway system. Cheap and available gasoline, cars, and highways spurred the development of suburban communities, many of which were financially supported by the government through the Veterans Administration (U.S. Department of Veterans Affairs) and the Federal Housing Administration's loans and tax incentives, developed to meet housing needs for returning soldiers and their families.

In the 1950s and 1960s, the petrochemical industry experienced a growth rate "two and a half times that of the gross national product."[8] Although some petrochemicals were produced from coal and petroleum during the interwar period, it was the investments and expansion of the major U.S. petroleum companies in the 1940s and 1950s that transformed the industry. As Peter Spitz remarks in a business history of petrochemicals, "Regardless of the fact that Europe's chemical industry was for a long time more advanced than that in the United States, the future of organic chemicals was going to be related to petroleum, not coal, as soon as companies such as Union Carbide, Standard Oil (New Jersey), Shell, and Dow turned their attention to the production of petrochemicals."[9] The vast supply of petroleum generated a cheap and ready source of petrochemicals. Chemical companies and oil producers began investing in value-added products made from petrochemicals, such as plastics, pharmaceuticals, and pesticides, and started to generate markets for these new products. In this way, supply drove demand and created new markets for goods. For example, large-scale direct foreign investments in agriculture in the 1960s created much-needed markets for Jersey Standard's pesticides.[10]

Consumer markets for plastics and pesticides emerged almost overnight. The new suburban garage or kitchen was a veritable chemistry lab, full of new plastic surfaces, new poisons, and new cleaning products. Most notorious, of course, was DDT, used during the Second World War for mosquito control in the Pacific and Mediterranean regions to protect American soldiers against typhus, malaria, and yellow fever. After the war, the pesticide became commercially available to local governments, communities, residents, and farmers, who liberally sprayed DDT in residential neighborhoods and households to control for mosquitoes and other "pests."[11] Many other pesticides, such as endrin, aldrin, and lindane, developed for war efforts, were rapidly introduced into large-scale agricultural production with the promise of raising crops yields and lowering food prices.

New plastics appeared on store shelves: the oven bag, a plastic bag that could be used to cook food at high temperatures; Saran Wrap, developed by Dow Chemical Company, as well as Reynold's polyethylene plastic films, created during the war and used commercially to encase precut meats and hold lunches for the on-the-go modern worker; plastic baby bottles for those long car trips; and Teflon pans that simplified cooking and cleaning. Plastic resins developed during the war to replace scarce tin as a preservative lining in cans (tinless cans) came into widespread use in the packaging of motor oil and food by the 1950s.[12]

By-products of these new products were chemical residues, which began to be found in food and milk supplies by the late 1940s. Low levels of DDT and other pesticides were detected in food and milk; DES, the synthetic estrogen given to animals to increase meat production, appeared in edible tissue; and chemical compounds used in plastic wraps and packaging migrated into cheese, meat, and other foods. The risks of chemical exposure were no longer confined to the industrial workplace or the war front. They appeared in homes and on the dinner table. As traces of pesticides and plastics were contaminating air, water, and soil, the issue of the safety of chemicals in food captured political attention in the early 1950s.

CHEMICALS IN FOOD IN THE 1950S

From 1950 to 1952, the House of Representatives formed a committee to investigate the impact of chemicals in food and held a series of hearings on the topic, led by Congressman James Delaney. A young liberal from the Queens District Attorney's office in New York with strong ties to the labor movement, Delaney first took his seat in Congress in 1944. Delaney's awareness about chemicals in the food supply began with DDT. Not long after the war, a fellow congressman told Delaney the following story. The congressman (whose name Delaney did not disclose) decided to spray his midwestern lakeshore property with DDT. Several days later, being an avid fisherman, he went out on the lake, only to find a disturbing number of dead fish floating in the water. Delaney found the story deeply unsettling

Figure 2. "Food products probers," January 11, 1952. *Front row, left to right:* Rep. Paul Jones (D-MO); Rep. James J. Delaney (D-NY), committee chairman; and Rep. Dr. Erland H. Hedrick (D-WV). *Back row, left to right:* Rep. Gordon L. McDonough (R-CA) and Rep. Thomas Abernathy (D-MI). (Photo credit: United Press/ACME.)

and shared it with the Speaker of the House, Sam Rayburn (D-TX). If DDT could kill the fish in a large lake, what were the risks to Americans' health, given that people were eating food sprayed with the pesticide, and given that the pesticide was showing up in cows' milk? Rayburn agreed to have Congress look into the issue. In 1950, he initiated the Select Committee to Investigate Chemicals in Food Production and appointed Delaney as its chair.[13]

As it turned out, Delaney's involvement proved to be critical. For the next several decades, he embraced issues related to chemical safety and the need for greater governmental oversight of chemical production. He championed new regulations in food and ultimately helped to pass what became his crowning achievement in public office: the 1958 Federal Food, Drug and Cosmetics Act, or the Food Additives Act. The act expanded federal oversight of chemicals in food and included a short clause, memorialized as the Delaney clause, that prohibited carcinogenic, or cancer-causing, chemicals from the food supply. Delaney spent much of his career defending this clause, which, as soon as it was passed into law, became a source of contentious ideological and scientific debate about the meaning of chemical safety.[14]

What troubled Congressman Delaney in the early 1950s was the tremendous lack of understanding about what new chemicals entering the food supply were doing to public health and the agricultural environment.[15] At the time of the hearings, chemicals were known to inadvertently contaminate food as they leached from new plastics used in food packaging, as pesticide spraying left residue on crops, or, in the case of DES, as the chemical migrated into animals' edible tissue. Chemical preservatives, coloring, and emulsifiers were also increasingly being intentionally added to foods to enhance their color and texture as well as to extend their shelf life. The scope of Delaney's hearing considered all manmade chemicals in food—whether their presence was intentional or unintentional. In 1952, Delaney's committee reported that the FDA listed 704 chemicals in use in food production, of which 428 were considered "safe."[16]

The paramount problem, Delaney asserted in an article in *American Magazine*, was the profound inadequacy of existing food law—"a tragic legal joker that permits us to become a nation of 150,000,000 guinea pigs guilelessly testing out chemicals that should have been tested adequately before they reached our kitchen shelves."[17] While his rhetoric was aimed at waking up the public, or perhaps frightening the housewife in order to build political support for reform, his fundamental premise was sound.[18] The FDA had only very limited authority under the 1938 Federal Food, Drug and Cosmetics Act (an amendment of the nation's first food safety law, the 1906 Pure Food and Drug Act) to set tolerance levels on "unavoidable poisonous substances," which included lead arsenate and Paris green, two inorganic pesticides. In the intervening years, the number of pesticides, preservatives, additives (including feed additives), and chemicals used in food packaging had exploded, as had their production levels, and there existed no requirement for testing these substances' safety before they came onto the market.

Food safety was an issue for which Delaney could find political support from his constituency in Long Island. It was a middle-class, consumer issue, and at the height of the cold war, when fear of nuclear war and the spread of communism strongly influenced foreign and domestic policies, it was also a politically moderate issue. Moreover, drawing consumer attention to "poisons" in the food supply did not collide with more politically sensitive issues of worker health and safety or, by association, the labor unions. Food safety was about consumer safety, and it affected congressmen and Long Island housewives alike. The target for legislative reform was the 1938 Federal Food, Drug and Cosmetics Act.

In 1906, rising concerns about food quality in the wake of reports about the unchecked adulteration of foods by substances of unknown risks or benefits had precipitated passage of the Progressive-era Pure Food and Drug Act. Publication of Upton Sinclair's *The Jungle* in 1905, which described in vivid detail the horrendous working conditions in the meatpacking industry, generated considerable public attention to the problem and the political motivation for the passage of the law. While the book was a strong critique of industrial capitalism, the Pure Food and Drug Act dealt with the safety of food per se, not the working conditions of industrial factories. The law prohibited foods with "any added poisonous or other added deleterious ingredient which may render such article injurious to health" and placed the burden of proving a substance was poisonous on a government agency: the Bureau of Chemistry, within the U.S. Department of Agriculture. This authority was later transferred to the Food, Drug and Insecticide Administration, created in 1927 and renamed the Food and Drug Administration in 1931.[19]

The first amendment to the law, the Federal Food, Drug and Cosmetics Act, which was passed as part of New Deal reforms in 1938, expanded the power of the FDA to regulate labeling on food, cosmetics, and drugs and to seize dangerous products. The political impetus for the bill's passage was the tragic case of Dr. Massengill's Elixir Sulfanilamide, a liquid form of a sulfa drug that contained a toxic solvent, which reportedly killed over hundred people.[20] "The policy tragedy of elixir sulfanilamide," as Daniel Carpenter explains, "established the basic lesson that undergirds gatekeeping power in American pharmaceutical policy. In the absence of a regulatory sentry at the border between drug development and market, this lesson says, people will be harmed, and massively so."[21]

As it related to the FDA's oversight of food safety, the 1938 law upheld the absolute restriction of poisons from the food supply, permitted the agency to develop lists of approved ingredients, and provided authority for seizure of unsafe foods. The law also extended regulatory authority over medical devices and cosmetics and required premarket approval for drug safety. The New Deal policy held that in the absence of regulatory oversight the market would not act alone to protect the public's health. The burden of demonstrating that a substance presented a "reasonable possibility of harm to consumers" fell onto the agency itself, not the regulated industry.[22] The 1938 law did not give the FDA the authority to require that a company notify the agency when it planned to use a new additive. Nor did it require that industry undertake toxicity testing. With the deluge of new petrochemicals in the late 1940s and 1950s, the agency was quickly overburdened. It could not require that companies provide it with any information, but at the same time it bore the heavy burden of ensuring that chemicals on the market would not harm consumers.

When Delaney's committee took up the issue of food safety in the early 1950s, they framed the problem very broadly and considered the potential impacts of chemicals in food on nutrition, soil quality, and animal and human health. The hearing's original mandate was to evaluate the "nature, extent, and effect of the use of chemicals, compounds, and synthetics in the production, processing, preparation, and packaging of food products" and their effects on the health of the nation and the agricultural economy. Further, the committee was to determine the "nature, extent, and effect" of fertilizers on soil, vegetation grown on that soil, animals eating that vegetation, the quality and quantity of food produced on such soil, and the health of the public.[23] In other words, Congress was to consider the environmental and health impacts of a petrochemical-based agricultural system. Yet after eight years of resistance and debate, the 1958 reform of the food law ultimately provided for the continued expansion of a highly industrialized food production system intimately dependent upon and integrated with petrochemical production.

In the process of determining how regulatory authority would or would not be expanded, Congress debated some fundamental political questions that would shape chemical debates for the next half century: Was exposure to pesticides such as DDT, or feed additives such as DES, or compounds in plastic wraps that leached into foods an acceptable risk to the general public? Was it necessary and unavoidable? How would the state, the legislature, and the responsible regulatory agency, in this case the FDA, protect public health while simultaneously supporting a national economy increasingly integrated with the petrochemical industry? Who bore the burden of determining safety and risk? How would safety be defined and determined? Did some chemicals present too great a risk to be permitted into American food, or could any industrial chemical be considered safe if it were present in small doses?

When Congress met to evaluate the safety of chemicals in food in 1950, it heard from many sectors of society, including scientists, nutritionists, agriculturalists, representatives of the FDA and the U.S. Public Health Service, pure food activists, and representatives of the National Canners

Association, the Grocery Manufacturers of America, the Manufacturing Chemists' Association (MCA), and the National Agricultural Chemicals Association.[24] The MCA, which represented U.S. chemical producers, together with the food and pesticide trade associations, delivered a resolute and unified message to the committee: any reform of the food laws or extension of the FDA's oversight of chemical safety was unnecessary. According to the chemical trade association, the 1938 Federal Food, Drug and Cosmetics Act sufficiently protected the public's health. In the early 1950s, regulatory reform was an unwelcome prospect for the food and chemical industries, which were just developing new commercial markets for pesticides, preservatives, and packaging products. The industry trade association representatives asserted to members of Congress that these new chemicals didn't represent significant or novel risks to the public's health.[25]

After all, "life has always been full of risk," quipped Dr. Robert Kehoe, director of the Kettering Laboratory at the University of Cincinnati, a research institution funded by the leaded-gasoline manufacturer Ethyl Corporation. From the 1920s to the 1950s, Kehoe's research on lead had provided the toxicological reasoning that supported the industry's position that lead exposure could be safe.[26] Relying on the same logic first used to defend the introduction of lead into gasoline in the 1920s, Kehoe reassured Congress that all chemical hazards could be safe. Sure, DDT could be found in milk, he remarked, but like lead, the pesticide did not present a risk to humans at the low levels to which humans were exposed. These chemicals were new and synthetically produced, but that did not necessarily make them more harmful than the chemicals all around us every day. Even natural chemicals were dangerous at high levels. Further, he argued, since pesticides were necessary to produce cheap food, some level of risk was inevitable and unavoidable. The key was "to get that concentration which is without harm and which, if it has benefit, will provide benefit without the effect of harm." The risk of chemical exposures was just part of modern life and the trade-off for the benefits such compounds brought to society and, of course, to their producing companies. The solution to this problem, he told members of Congress, would be found, not in regulatory mechanisms to limit chemicals, but rather in expanded research to ensure their safe use. In other words, science would determine the levels at which humans could be safely exposed to all chemical hazards.[27]

Charles W. Crawford, commissioner of the FDA from 1951 to 1954, strongly disagreed with the industry's confidence in the existing law. He had worked for the regulatory agency for thirty-four years and knew well the limits of the law given recent changes in food production. While the 1938 Food, Drug and Cosmetics Act theoretically restricted poisonous substances from the food supply, he argued, it provided no sufficient definition of or process to determine the safety of chemicals.[28] Indeed, the lack of any safety standards in the 1938 law had been a major disappointment for some Progressive reformers at the time. The assistant secretary of agriculture under President Franklin D. Roosevelt, former Columbia University professor Rex Tugwell, who fought to lower pesticide tolerance levels and restrict dangerous chemicals from cosmetics, considered the 1938 law "disgraceful" because it set no standards, no penalties for fraud, and no restrictions on dangerous products and because it required testing only for drugs.[29] The failure of the existing law was precisely why some companies, such as Beech-Nut, were forced to establish safety standards for themselves. Testifying before Congress in 1950, Commissioner Crawford roundly supported the requirement of premarket safety testing for all chemicals added to food, either directly, as in the case of preservatives and dyes, or indirectly, as in the case of chemicals that inadvertently contaminated food, such as compounds that leached out of resin in plastic packaging.[30]

Determining what was safe was not as simple as Kehoe's argument that one must find the concentration where benefits outweigh risks. Awareness of the complexity of chemical risks was implicit in the Beech-Nut representative's concern for the potential long-term reproductive effects of pesticides in children. If one followed Kehoe's simple logic, a chemical's safety need not be based on who was exposed or how the chemical behaved. All toxic chemicals were presumed to function similarly in the complex process of human life and development, regardless of when exposure occurred. However, as the example of the estrogenic effects of DDT made clear in the 1950s, not all chemicals behaved similarly or simply. Most importantly, toxicity and carcinogenicity were not the same phenomenon.

Dr. Wilhelm Hueper, director of the Environmental Cancer Section of the National Cancer Institute (NCI), explained to members of Congress the difference between toxic chemicals and chemicals that cause cancer. Some compounds demonstrate very little toxicity, in that only very high doses of them result in death, but they are nonetheless highly carcinogenic. In other words, a chemical at a certain concentration might not cause immediate mortality, but over the long term it could cause cancer. To say that a chemical is nontoxic or has low toxicity, therefore, does not necessarily translate into noncarcinogenicity. Toxicity and carcinogenicity, Hueper argued, are not interchangeable categories, but rather separate biological processes. This distinction, according to Hueper, had been demonstrated in studies of reproductive cancers from DES exposure, as well as the carcinogenic effects of beta-naphthylamine. Both substances demonstrated low toxicity and carcinogenicity. Hueper stated that chemical safety tests should require consideration of the complexity of carcinogenicity, because of cancer's long latency period and evidence of carcinogens' effects on the young via maternal exposure. He recommended to Congress "that the uncontrolled use of any known or suspected agent with carcinogenic properties is not advisable."[31]

Hueper's testimony before Congress, like his long research career, was highly controversial. A leader in the field of chemical carcinogens, Hueper was appointed as the first and, as it turned out, the only director of the NCI's Environmental Cancer Section in 1948. (The NCI was established in 1937 as part of the National Institutes of Health [NIH].) He had been fired from his position as chief pathologist at the DuPont Company in 1937 after he reported bladder cancers in workers exposed to the company's naphthylamine dye. For years, the company attempted to discredit Hueper's reputation, and it barred him from publishing or presenting any of his research findings from the company's lab. Dr. G.H. Gehrmann, a DuPont scientist, sent false charges to the Federal Bureau of Investigation (FBI) that Hueper had been a member of the Nazi Party. (Hueper had come to the United States from Germany in the 1930s.)[32] Undeterred, Hueper published the first extensive review of chemical carcinogens in the workplace, *Occupational Tumors and Allied Diseases*, in 1942.[33] Rachel Carson drew heavily from that text in her research for *Silent Spring* and spoke with Hueper several times while writing the book. In *Silent Spring*, she repeats Hueper's warning that exposure to carcinogens during pregnancy can result in cancers in the young.[34]

Hueper served as director of the NCI's Environmental Cancer Section until 1964, but his tenure was plagued by persistent efforts to restrict his work, travel, and presentations, including testimony before Congress regarding the safety of chemicals in the food supply in the 1950s. In a twenty-page memo to the director of the NCI in 1959, he detailed attempts to discredit and slander his research and reputation, including efforts by senior staff at the FDA to block his testimony at the food additive hearings in the late 1950s by threatening to revoke his promotion.[35] In 1958, Leroy Edgar Burney, U.S. surgeon general, denied the publication of a lengthy paper written by Hueper on the health hazards of food additives in consumer products, deeming it "unsuitable" at a time when Congress was debating reform of the food law.[36] Dr. John Harvey, deputy commissioner of the FDA, told a Washington newspaper that Hueper's paper would raise unnecessary alarm in the public.[37]

While Hueper's views on chemical carcinogens were unpopular and Hueper himself a controversial man, he was not alone in his concerns. His colleague Dr. William Smith, a former professor of industrial medicine at New York University (NYU) and chair of the International Union against Cancer, also came before Congress in the 1950s to explain the risks of chemical carcinogens. Like Hueper, Smith allegedly had lost his job at NYU because of his research on chemical hazards in the workplace. Smith informed members of Congress that pressure from the chemical industry had successfully halted research programs on environmental cancer at the NCI as well as at NYU. Industry pressure, Smith alleged, stymied research on chemical carcinogens by reducing available funding and eroding institutional and reputational support for the few researchers in the field. The NCI and NYU denied such accusations, but Hueper defended them.[38]

The International Union against Cancer was an affiliation of cancer organizations from fifty nations, among them the U.S. Public Health Service. The organization first took up the issue of chemicals in food in 1939, but the outbreak of the Second World War delayed its activities until 1950, when the union reconvened and established a "unified position against cancer-causing chemicals in food." Four years later, well aware of the efforts to reform the food law in the United States, the union held its fifth meeting to discuss the problem of what it called "irreversible" toxic chemicals: those chemicals for which the union assumed no threshold of safe exposure existed, namely cancer-causing compounds or carcinogens. In 1956, the union issued a statement that called on all nations to restrict carcinogens from the food supply.[39]

CARCINOGENS IN FOOD

Of particular concern to both Hueper and Smith was the use of the potent synthetic estrogen DES in livestock. DES was the first synthetic (non-steroidal) form of estrogen, developed by the British biochemist Edward Charles Dodds in the mid-1930s and commercialized in the early 1940s to treat numerous female "problems": menstruation difficulties, menopause symptoms, nausea during pregnancy, prevention of miscarriages. The drug also found use as a feed additive in poultry and livestock, as it increased the animals' weight and thus meat production.[40]

Estrogen was first isolated in the late 1920s. The isolation and identification of hormones, which were then called internal secretions, including estrogen, testosterone, and insulin, were believed to hold great promise for the treatment of chronic diseases. The most promising discovery was insulin in the early 1920s: the hormone could bring diabetic patients back from the brink of death. Moreover, the ability to produce synthetic forms of insulin expanded medical treatment for hundreds. In the hopes of finding more wonder drugs, Edward Doisy, a biochemist who participated in the research to isolate insulin, worked to isolate estrogen with a fellow colleague, Edgar Allen, a biologist studying estrus, or menstrual cycles, in mice, at Washington University School of Medicine in St. Louis. They succeeded in 1929.

While estrogen had been studied by researchers as an elixir of youth and fertility and as a chemical messenger of femininity (as well as masculinity) as early as the mid-nineteenth century, its chemical isolation provided a pathway for developing cheaper, synthetic forms of the hormone, like DES, and increasing the market for the drug. The emerging market for estrogens involved the treatment of female reproductive issues, such as menopause, menstrual irregularities, miscarriage, and female behavior associated with menstruation (premenstrual syndrome), which were increasingly medicalized in the early twentieth century.[41]

Edward Charles Dodds studied and then taught in the small biochemistry department at the University of London Middlesex Medical School in the 1920s and 1930s. He became fascinated with estrogen and its powerful ability to cause cells to proliferate. He presciently noted, "It is justifiable to say that hitherto no body has been isolated with such a powerful action upon the growth of cells, and it will not be surprising if this and similar bodies start a new era in the investigation of cell growth, and possibly of malignant disease."[42] Dodds began his work in search of a less expensive synthetic form of the hormone in the late 1920s.

Contrary to the prevailing paradigm in hormone research at the time, Dodds and several of his colleagues believed that many chemical structures were capable of estrogenic activity. His research sought to identify the structure that provided the ultimate source of estrogen's activity. He did this by testing a number of compounds with chemical structures similar to that of natural estrogen. To test whether the chemical compounds exhibited estrogenic properties, Dodds removed the ovaries—the source of estrogen—of rodents and then exposed the animals to the various compounds. Removing the ovaries stopped estrus in the animals. Dodds identified a chemical as estrogenic if it induced estrus or caused the cornification of vaginal cells (the cells turned to harder tissue). The stronger the estrogen, the lower the concentration needed to elicit any of these responses.[43]

Dodds tested a number of chemicals throughout the 1930s, including polycyclic aromatic compounds, identified as chemical carcinogens by his colleague James Cook, and fifteen derivatives of diphenyl and diphenyl methane (compounds with two phenol rings).[44] Among the derivatives tested was BPA. In a 1936 paper, published in *Nature*, Dodds reported BPA's positive estrogenic response.[45] Two years later, he published a paper that identified what he believed was the "mother-substance" of all estrogenic compounds: DES.[46] The highly active estrogenicity of DES (only a low dose was needed to induce estrus in ovarectomized rodents), its purported low toxicity, and the fact that it was fully active when given orally to humans made it an ideal therapeutic drug, Dodds argued in the early 1950s.[47]

DES and Premarin, a mixture of natural estrogens, were the first drugs reviewed by the FDA after the passage of the 1938 Federal Food, Drug and Cosmetics Act. The approval process for Premarin was fairly simple and uncontested, given that a number of natural estrogens were already on the market. DES, on the other hand, was a novel drug with unproven therapeutic uses and high estrogenic potency. In *Toxic Bodies*, historian Nancy Langston provides a detailed account of the FDA's approval of DES in the 1940s despite concerns about its potential carcinogenicity and reproductive toxicity. Research on DES in the 1930s demonstrated reproductive toxicity in laboratory animals and generally raised ongoing questions about the carcinogenicity of estrogen, whether natural or synthetic. Further, DES's therapeutic possibilities were unclear and undemonstrated, with reported side effects in women, such as nausea and vomiting. The FDA initially instructed drug companies to withdraw their applications for DES's drug approval. Yet deference to physicians, who largely supported the drug, combined with coordinated political pressure from industry, contributed to the drug's final approval in 1941.[48] Langston explores a number of additional reasons for the agency's decision, including recognition by the FDA that some of its skepticism about the drug's safety was based "on evidence that could not be defended in court and therefore could not stand up against political pressure."[49]

Hueper and Smith were particularly concerned with the public's exposure to the low levels of DES used to increased meat production in poultry and other livestock. The FDA allowed the use of DES in poultry in 1947 despite evidence of its residues in edible portions of meat, and use continued in the 1950s even after reports of reproductive effects in male restaurant workers and minks that consumed chicken heads (where the DES was implanted). In 1954, researchers at Iowa State College identified DES's ability to dramatically increase weight gain in cattle, and, although the FDA ultimately banned its use in poultry in 1959 due to its persistent detection in tissue, the FDA allowed its continued use in cattle until the end of the 1970s, given ongoing affirmations by producers that no DES was detectable in meat. Indeed, Eli Lilly and Company, a producer of DES for human treatment, successfully obtained agency approval for DES use in cattle because it demonstrated that no amount could be detected in the edible tissue of animals.[50]

In 1956, Smith, along with Granville F. Knight, Martin Coda (the president and vice president, respectively, of the American Academy of Nutrition), and Rigoberto Iglesias, a physician and researcher from Chile, presented a paper on the dangers of DES at an FDA meeting on medicated feeds. They declared that DES was "known to induce cancer" and listed a number of reproductive pathologies associated with very low doses of DES—levels at which "the effective dose approaches the infinitesimal." They conjectured that the "administration of estrogens, among which diethylstilbestrol is one of the most potent, has led to a wide range of pathological changes in human beings and in animals. In mice, rats or guinea pigs, estrogens can induce polyps, fibroids and cancers of the uterus, cancers of the cervix, cancers of the breast, hyperplasia of prostatic stroma and of endometrium, tumors of the testicle and hypophysis, and lympho-sarcomas."[51] Such carcinogenic effects, according to these scientists, not only could occur at extremely low doses but also could result from continuous exposure. They wrote: "Most important, it has been learned from animal work that intermittent injection of very large doses of estrogens is far less effective in inducing tumors than is a *continuing* exposure to an extremely minute dose. This phenomenon has been repeatedly observed by one of us (R.I.) [Rigoberto Iglesias] in experiments conducted over a period of nearly twenty years. It is a *continuing* exposure to extremely minute doses that is to be feared from the introduction of estrogens into the food supply."[52]

Such concerns were a minority view at this meeting of regulatory agencies and industry representatives. Speakers from Lilly, Pfizer, and the Department of Agriculture questioned the legitimacy of such damaging DES research by highlighting common exposure to natural estrogens. Naturally occurring estrogens are present in foods and the human body, which meant, the speakers argued, that they don't necessarily present any harm. Moreover, the fact that DES had some adverse effects in one species didn't necessarily mean they would occur in humans. (Such efforts to minimize or generate uncertainty about any potential risk of synthetic or environmental estrogens would become a common tactic in attempts to challenge the validity of endocrine disruption forty years later.) For drug companies and meat producers in the mid-1950s, the benefits of increasing meat supply and thus making larger profits were far too great to worry about long-term exposure to minute amounts of the drug. President Eisenhower sent a message to the meeting, calling "the extensive and increasing use of medicated feeds supplied by a great and growing industry . . . a boon to small scale as well as large producers of livestock," to the benefit of all.[53]

For Smith and his colleagues, the fact that human exposure to carcinogens in the food supply mirrored the conditions of their experiments—low doses and continuous exposures—raised serious concern about the wisdom of widespread use of DES in food production. Fueling their alarm was the long latency period of cancer, which meant that effects of exposure to these chemicals wouldn't be detected for decades to come; by that point, the adverse effects would manifest and prevention would be too late. Smith and Hueper also worried about the recognized vulnerability of the young. As Hueper explained to Congress, "The newborn appears more sensitive to small doses of carcinogenic chemicals [than the adult]."[54]

Hueper's concerns about low levels of chemical carcinogens also grew out of his research on several new plastics increasingly being used to package food. During Delaney's hearings, members of Congress heard about the expanding prevalence of new plastic containers and wraps.[55] Many of these plastics could contaminate food because they leached chemical components of their polymers (i.e., complex chemical compounds composed of repeating molecular structures, which include proteins, DNA, carbohydrates, and plastics). Don Irish, a representative from Dow Chemical Company, confirmed before Congress that "in many types of plastics there are small amounts of ingredients which can be extracted" and thus released into food, making it necessary in these cases, he asserted, to determine their level of safety.[56] In response to known leaching from plastics wraps, the Department of Agriculture, in the early 1950s, sought to require safety testing by the food industry. The government demanded that the tests examine health effects over a long duration of use, whereas the food industry contended that short-duration tests were sufficient.[57] Expanding regulatory oversight of substances used in plastics packaging was not a welcome possibility for manufacturers of chemicals or plastics. Regulatory oversight was a new barrier to potential markets for plastics and extensive testing introduced new costs to business. Yet given emerging evidence of the carcinogenic effects of several plastics, there was reason to be concerned.

Barred from doing occupational research by officials at the NIH, Hueper conducted several experimental studies of plastics using rodents. In 1956, he published a study that reported carcinogenic responses in rats exposed to polyvinylpyrrolidone (PVP), a polymer used for diverse purposes, including as an adhesive and emulsifier (a substance that combines two other substances that would not otherwise mix). The following year, another researcher reproduced Hueper's findings, further substantiating PVP's carcinogenicity.[58]

Not all research on plastics began with an intentional design. Some resulted from moments of serendipity in the research laboratory as plastics expanded into new markets. Plastic wraps, like plastic containers in the 1980s and 1990s, were commonly used in the kitchen, as well as scientific laboratories, by the 1950s. In the early 1950s, while conducting a study of a hypertension drug in rats, Dr. Bernard Oppenheimer, a cardiologist, and Dr. Enid Oppenheimer, a physiologist (along with their colleague Arthur Purdy Stout), researchers at the College of Physicians and Surgeons at Columbia University, wrapped rat kidneys in plastic film to induce high blood pressure in the animals. After several years, the Columbia researchers observed that seven of these same rats had developed malignant tumors at the sites where their kidneys were encased in the film and that the cancer had metastasized to other organs. The researchers also found that tumors could be induced if they imbedded plastic film in the abdominal wall of rats. These findings, the researchers remarked, led them to investigate which chemicals might be causing the tumors.[59] In *Cancer Research* in 1955, the researchers described the very high rate of malignant tumors that developed over several years in rats exposed to a number of plastics, including polyvinyl chloride (PVC), Saran Wrap, polyethylene, Dacron, cellophane, and Teflon.[60] With funding from the NCI, the Oppenheimer team began a series of investigations into plastic films and the monomers added to plastics as stabilizers, plasticizers, and catalysts.[61]

News of these findings received national attention. "The discovery by Dr. Oppenheimer and his associates is important because it indicates that there [are] probably scores of plastics that can generate tumors," the *New York Times* reported in 1952.[62] News that hundreds of chemicals were making their way into food without any testing provided fodder for reform. If Hueper, Smith, and his colleagues' concerns were credible, the presence of carcinogens even at low levels could present a serious long-term public health risk. Such scientific reasoning lent legitimacy to Delaney's support for regulatory oversight of *all* chemicals used in food processing, production, and packaging. Delaney's committee drew up a long list of chemicals that should be subjected to regulatory oversight, including pesticides, insecticides, growth regulators, preservatives, and mold inhibitors, as well as chemicals used in plastic packaging, including resins and plasticizers used in tin-can coatings (i.e., phenol-formaldehyde resins and phthalates), for which there was little toxicological information.[63] The committee's report brought the issue to national attention, and for the next six years Congress considered efforts to reform the Food, Drug and Cosmetics Act.[64] A first step in this process was the Miller Amendment to the act, passed in 1954, which established tolerance levels for pesticide residues in food. Delaney considered the amendment to be only a partial response to his committee's full recommendation: the premarket testing of *all* chemicals, not just pesticides, contaminating food.

REGULATING CHEMICALS IN THE FOOD SUPPLY

Delaney's hearings in the early 1950s galvanized a nascent consumer movement. In 1955, a small consumer health advocacy group, the National Health Federation, formed to apply direct pressure on Congress to reform food legislation. It initiated a letter-writing campaign aimed at members of Congress calling for reform; dozens of letters poured into Delaney's office from citizens across the country, encouraging the congressman to initiate additional hearings on food safety. The letters frequently included a request for a copy of the 1950–52 hearings. Indeed, the Government Printing Office received so many requests that it had to print additional copies of the hearings.[65]

In 1956, Gloria Swanson, an Academy Award–winning actress, a fashion icon of the 1920s and 1930s and an avid health food advocate, addressed a gathering of congressmen's wives in Washington on the need for food law reform. (Unlike today, most congressmen moved their entire families to Washington during their terms.) Delaney recalled that immediately after Swanson's address members of Congress approached him in the halls to inform him that their wives insisted that they support his legislation. After nearly losing his seat because of what he considered to be industry pressure, Delaney returned to Capitol Hill in 1956 and initiated a second round of hearings to discuss a number of new reform bills.[66]

In response to increased public and political pressure, the chemical industry's strategy shifted significantly in the mid-1950s. Its initial opposition to reform, voiced at Delaney's first hearings, gave way to industry-sponsored legislation. After Delaney's committee released its recommendations for premarket testing of all chemicals in food production in 1952, the chemical trade association MCA hired its first public relations firm, Hill and Knowlton, and registered for the first time as a lobbying organization. By the early 1950s, the chemical industry, according to MCA president George Merck, had yet to spend even two hundred thousand dollars on public relations. This was in stark contrast to the steel and petroleum industries, which at the time each spent around two million dollars a year. In a discussion of this new public relations push, Merck, chairman of Merck and Company, "enjoined the chemical industry to associate themselves with 'science' in the public mind." He explained, "[The] public has been taught to be interested in science; moreover, they have come to respect the scientist. The industry would do well to identify itself to the public as the organizational mechanism through which the contributions of science are made available for everyday use."[67] In other words, the trade association should speak for science in order to instill public trust in the industry and in the safety and benefits of its products.

The struggle to manage and directly influence the legislative reform process was the chemical and plastics trade associations' first significant battle in the decades before major environmental and occupational safety and health laws were passed. It was in the 1950s that the MCA began transforming into the organization that the leading lawyer for the plastics industry, Jerome Heckman, would later call a "quasi-public institution": one that speaks for science and advises on the direction and implementation of regulatory laws.[68]

Perhaps the greatest threat to industry trade representatives in negotiations over food legislation was the position of the International Union against Cancer on carcinogens. In a lengthy letter to Delaney in 1957, Smith stressed the special dangers of carcinogens and recommended additional legislative language to address these risks. "Responsible bodies of experts," including the International Union against Cancer, Smith wrote, "have clearly stated that safe doses cannot yet be established with certainty for carcinogens." Smith explained that in his own work with laboratory animals he had induced cancerous tumors with "single, very small, doses" of known carcinogens. Working with the chemical urethane, Smith continued, he had induced lung tumors by exposing female rats during their pregnancies. Without knowledge of the safe dose of carcinogens, and given the consensus among international experts that small doses could induce cancer, Smith forcefully argued against allowing carcinogens into the food supply. There is "no moral justification," Smith wrote, "for obliging the consumer, many of whom are children, with life expectancies serving the long latent period of carcinogens, to accept such needless

risks."[69] For him the problem was simple: the risks were too high—small doses of carcinogens could cause cancer, and these hazards were in the food supply—and carcinogens' benefits were questionable. Given the grave risks of carcinogenic substances, Smith recommended that the following statement be included in the new law: "The Secretary shall in no case approve for use in food any substance found to induce cancer in man or in tests upon animals."[70] According to Smith, some chemicals should be considered unacceptable regardless of the concentration at which they are present in the food—that is, they should be considered unsafe per se.

Faced with inevitable reform of the food law and demands to restrict all carcinogens from the food supply, the chemical and plastics trade associations changed tactics, and by 1957 sought to develop and promote legislation that would provide "the most comprehensive exemption provision possible."[71] If reform was coming, the industry trade associations wanted to ensure that it would be the reform least burdensome to business. This meant arguing for the absolute necessity of hundreds of chemicals, including chemical carcinogens, in food production and limiting federal oversight and regulation of these compounds, while at the same time providing the legal security of federally sanctioned safety standards.

DEFINING SAFETY IN THE FOOD ADDITIVES ACT

The shift in industry support for food reform may also be attributed to the new leadership at the FDA, Commissioner George Larrick, who began his long career at the FDA in 1923 as a food and drug inspector and served as commissioner after Charles Crawford, from 1954 to 1962. According to the 1938 Food, Drug and Cosmetic Act, all poisonous substances were to be absolutely restricted from food. Such an antiquated approach to chemicals, Larrick told members of Congress, failed to account for profound changes in food production since 1938. Risks of exposure were inevitable and unavoidable aspects of modern life. Therefore, any attempt to absolutely restrict chemicals, he contended, was naive and impractical.[72] His position was welcome news to industry. Defining chemical hazards as poisons per se that must be banned from food was an impossible position from industry's perspective.

During Larrick's tenure, the FDA's budget increased tenfold, followed by expansions in facilities, laboratories, and regulatory oversight.[73] In Larrick, the drug and chemical industries found a commissioner who shared their policy positions, or at least could be influenced to do so. As Daniel Carpenter remarks in his history of drug policy at the FDA,

"Larrick is portrayed alternatively as a bumble or an industry sop. Images abound of drug company officials walking unimpeded through the agency's corridors, setting up informal residence and hounding medical reviewers."[74]

When asked by Congressman Joseph O'Hara (R-MN), who worked with the MCA to introduce a food reform bill, whether he believed that food manufacturers were concerned with protecting the public interest, Larrick answered, "Yes." This new attitude toward industry, O'Hara remarked, provided "great relief." Reversal of the 1938 "strict policy of the law against poisons in our food" could be done safely, Larrick explained, sounding not unlike Robert Kehoe in the early 1950s, by minimizing public exposure to chemical hazards. In this way, the purported benefits of chemicals could be reaped while protecting public health.[75]

Lawrence Coleman of the MCA echoed Larrick's articulation of chemical safety. "We must divorce from our thinking and from the administration of the food laws," Coleman contended in the hearings, "the idea that 'poisonous' or 'deleterious' are absolute concepts; that a food additive can be denominated a poison without reference to the quantity ingested by a human being." Poison, he argued, was a "relative concept."[76] In other words, "Dosage is key" to determining safety.[77]

The notion that a chemical's toxicity was relative to its dose followed the legal principle of *de minimus non curat lex*, or "the law does not concern itself with trifles."[78] That is, at the very low concentrations where toxic effects declined, the law might not apply. A *de minimus* approach to chemical hazards contrasted directly with the concept of a chemical hazard or poison per se, as articulated in the 1938 law. Under a *de minimus* interpretation, a chemical could be safely permitted in food when the risks were minimal. This shift in the legal framework of food law was essential for industry support of reform.

And, indeed, the final version of the food reform legislation, the 1958 Federal Food, Drug and Cosmetics Act, or the Food Additive Act, included a regulatory framework quite favorable to the chemical trade association. While the law did require premarket testing for all chemicals in food, as called for by Delaney's committee, it also included a number of loopholes for excluding chemicals from regulatory oversight. These included those compounds in use before the amended law, chemicals regarded as safe, and those below any detectable level in food. Such exemptions are evident in the definition of a "food additive" under the 1958 food law.

The law defines a food additive as "any substance the intended use of which results or may reasonably be expected to result, directly or indirectly, in its becoming a component or otherwise affecting the characteristics of any food."[79] The phrase "may reasonably be expected" provided for regulation that followed the *de minimus* standard. In other words, a chemical had to "reasonably be" present in food to be deemed a food additive. Substances that were "generally recognized, among experts . . . to be safe under the conditions of its intended use" would be exempted.[80] This "generally recognized as safe" clause in the law, which became known as the GRAS standard, grandfathered in hundreds of chemicals used before 1958 as *prima facie* safe, requiring no testing and marking a tremendous victory for industry.[81]

The 1958 law also included a judicial review clause, which the chemical industry strongly supported. Judicial review provided a legal process through which FDA decisions and actions could be challenged in the courts. In the 1950s, the judicial review process established a procedural check on the administrative authority of regulatory agencies. (Twenty years later, environmental organizations began to successfully use this system of checks and balances to push for regulatory oversight and stronger safety standards.)

Legal authority for the judicial review clause had been only recently established under the Administrative Procedures Act (APA), passed by Congress in 1946.[82] Under the APA, the courts are required to review any regulatory action to determine whether it conforms to the original statute and if it is "arbitrary, capricious, an abuse of discretion or otherwise not in accordance with the law." In 1951, the Supreme Court ruled, in *Universal Camera Corporation v. National Labor Relations Board*, that the APA directed the courts to take greater responsibility in determining the "reasonableness" and "fairness" of regulatory decisions. The *Universal* decision ruled that in evaluating "reasonableness" the courts must consider the "weight of the evidence," including "whatever in the record fairly detracts from the weight of the evidence."[83]

Support of judicial review came from the American Bar Association and conservative political leaders seeking to place checks on the expanded regulatory authority promulgated by New Deal legislation.[84] And, indeed, the FDA recognized judicial review as a threat to its expert opinion and autonomy. William Goodrich, assistant general counsel of the FDA, came before members of Congress during a food reform hearing to vehemently oppose such a provision. Throwing open expert opinion and agency decision making to the courts, Goodrich argued, ultimately would shift the burden of proof from the industries to the FDA. In court, the agency would have to prove the *lack* of safety, as opposed to industry having to prove safety. Defining safety, Goodrich contended, should not be "a popularity contest before the judge. It is not a question of demeanor or veracity. . . . It is an informed, understanding study of a scientific experiment. I don't believe that juries or judges . . . are competent to pass on that kind of question."[85]

Goodrich made a subtle point worth highlighting as the courts today play increasingly powerful roles as arbiters of scientific disputes in regulations. The regulated industry welcomed the burden of testing, or determining safety, as a form of self-regulation.[86] While this shifting of the burden of testing from the FDA to industry was necessary and welcomed by the agency, providing industry with the legal means to challenge a scientific decision by the agency effectively moved the task of demonstrating lack of safety or a need for regulatory action back to the FDA. In this way, industry effectively controlled safety determination through either self-regulation or the courts. The result, from Goodrich's perspective, was a significant undermining of agency expertise and authority. His concerns were prescient. In the years to come, efforts by the FDA and later the EPA to set regulatory safety standards were chronically stalled and delayed by lengthy judicial reviews.

Finally, industry and probusiness legislators also successfully defeated the FDA's demand that producers define the "functional value" of a food additive, a requirement supported by Larrick. This would have created regulatory oversight of the benefits of a chemical and its impact on nutrition. A food additive would then need to be both safe and beneficial. At the time, the FDA lacked the authority to demand safety and effectiveness of drugs, a provision that would come in 1962 with the Kefauver-Harris Amendment. Conservative lawmakers viewed such a provision as an affront to the free market. In questioning the commissioner, Congressman Martin Dies (D-TX) questioned whether, in a "free enterprise" society, the government should make decisions about what consumers consider beneficial to their lives: "The best criteria is whether the public is willing to buy the product. They are the final arbiters," not the FDA, Dies argued to Larrick.[87]

From the perspective of the affected industries, the GRAS clause, the reversal of the FDA's policy to absolutely restrict chemicals as "poisons," the inclusion of the judicial review process, and the cutting of the functional value clause added up to a very favorable law. Safety would be determined according to dose, and the "reasonableness" of the standard established could be challenged in court if necessary. Industry might have claimed a wholesale victory, save for one clause that created an exception for carcinogens.

In the final negotiations on the bill, Delaney lost considerable ground to industry demands and FDA concessions, and Smith condemned the FDA for its failure to support Delaney's bill: "I am told . . . that FDA prepared that bill with the knowledge and approval of food industries." He went on to decry the bill's failure to address carcinogens.[88] The absence of Smith's language on carcinogens in the final version of the bill provoked Delaney to stall its passage until Larrick agreed to meet with the congressmen to discuss a possible compromise. After a closed-door negotiation with the FDA, the agency agreed to support inclusion of the following passage: "No additive shall be deemed to be safe if it is found to induce cancer when ingested by man or animal, or if it is found, after tests which are appropriate for the evaluation of the safety of food additives, to induce cancer in man or animals."[89] This language became known as the Delaney clause, the anticancer clause, or the zero-tolerance clause. It opened the door for a per se ruling on those chemicals deemed carcinogenic and established a legal trigger to ban substances. The prevailing question for industry trade lawyers upon the passage of the 1958 Food, Drug and Cosmetics Act was how the FDA would interpret the Delaney clause. It quickly became a significant thorn in the side of the chemical industry and its lawyers, who would work for decades to minimize its impact and ultimately to remove it from parts of the law.

The obvious problem for the chemical trade association was that chemical carcinogens—DES, plastics, and some pesticides—were used in food production. For example, in February 1957, the MCA's Plastics Committee discussed a number of recently published papers, including the Oppenheimers' work, on the carcinogenic effects of plastic films in food packaging.[90] "The Committee reviewed recent allegations that some plastic materials and specifically polyethylene might promote cancerous growth. In recognition of the importance of plastic materials in food handling and the damaging effect of such irresponsible claims, the Committee asked for the advice of the MCA's Chemicals in Foods Committee as to what should be done, if anything."[91] A review of the literature on toxicity of plastics was sponsored by the B.F. Goodrich Company. The report, presented at the Industrial Hygiene Conference in 1959, cleared many plastics as safe for human use. In summaries of nylon, polyethylene, and vinyl resins, the B.F. Goodrich researchers referenced the work of the Oppenheimers and Hueper, yet concluded that all these plastics were "nontoxic," making no mention of carcinogenicity.[92]

IMPLEMENTING THE FOOD ADDITIVES ACT

One lawyer in particular played an essential role in implementing the new food law: Jerome Heckman (see the introduction), a lawyer for the Society of the Plastics Industry (SPI), the trade association of the plastics industry. Heckman effectively guided those in the chemical industry affected by the new law through the loopholes of the regulations and skillfully interpreted the Delaney clause. In the process, he built a global law firm with leading expertise in food law.

A 1952 graduate of Georgetown Law School, Heckman took his first job with the law firm Dow, Lohnen and Albertson, with an interest in radio communications. In one of Heckman's first cases, he represented George Russell, a prominent businessman who owned several radio and television stations and the leading newspaper in Seattle, and who was "active in the lumber industry." Russell had developed a company that manufactured equipment for laminating "wood shreds together with glue by passing the combination through a wave guide where radio energy" melted the glue. The company faced a serious obstacle in using the technology because the FCC restricted many uses of radio frequencies. Working with Russell, Heckman became an expert in the technology and made recommendations to the FCC that resulted in rulings in the client's favor. As it turned out, the plastics industry also used this technology and was similarly concerned about the FCC rules. This brought the SPI's executive vice president to Dow, Lohnen and Albertson and subsequently to Heckman, who worked on the trade association's FCC case. The SPI ultimately retained Heckman, still a young but accomplished associate. Heckman became the SPI's general counsel, a position he holds to this day at his law firm, Keller and Heckman.[93] Heckman's success in keeping markets open for the plastics industry earned him godlike status at the trade association and a place in the Plastics Hall of Fame.[94]

In the years immediately following passage of the food law, Heckman worked with the FDA and SPI to implement the law in a way that would best support his clients' interests. In his regular meetings with the FDA in the late 1950s and early 1960s, Heckman ascertained the agency's position on the Delaney clause. He recognized that any public argument against the need to protect Americans from carcinogens was an impossible one for the industry to make. As Heckman later remarked, "How do you argue against that?!" What Heckman sought and received from the agency was the assurance that the FDA was not planning to enforce the Delaney clause as a per se standard. The carcinogen clause, Heckman informed industry leaders, would be interpreted in accordance with the concept that risk was relative to dose. No carcinogens allowed in food, therefore, would be interpreted, not as absolute, theoretical zero, but as no detectable amount.[95] If the agency were to have interpreted the clause as a restriction on carcinogens per se (i.e., theoretical zero), it would have led to the banning of nearly everything, Heckman noted many years later.[96]

He also worked to secure legal exemptions for those chemicals in the GRAS category and those in "prior use." The GRAS category included over two hundred chemicals that were considered safe by "experts qualified by scientific training" and thus would not require additional safety testing. The second exemption, prior use, included those substances that had been sanctioned by the FDA before 1958. This category included nearly every major plastic product in use before 1958, including vinyl chloride.[97]

For those chemicals that still needed FDA approval, Heckman worked with the agency to establish a mutually agreed-upon process to submit information necessary for market approval (i.e., petitions) to the agency. An important aspect was minimization of requirements for submitting toxicological and migration data to the FDA. Petitions, according to Heckman, needed to follow a standard procedure that would protect the collective industry, as opposed to providing a single manufacturer with a competitive edge. Inconsistency in detection limits used by producers and differences in migration levels demanded standardization of methodologies. For example, the agency reported to the industry that polyethylene was found in fatty foods up to 100 parts per million (ppm), whereas the prior sanctioned use had allowed only 1 to 2 ppm. Polyethylene and polystyrene manufacturers, organized by the SPI, responded by developing standardized methods to detect migration and to determine how a petition would be written and submitted to the FDA.[98]

Speaking before the SPI's Food Packaging Materials Committee, Heckman informed producers that petitions did not require the submission of any toxicological data, on the basis of the "fact that most of the chemical substances involved are either prior sanctioned, or generally recognized as safe, or that toxicity data has previously been submitted in filing with Food and Drug by individual companies." Manufacturers should provide methods for determining migration levels but did not need to file specific analytical data. The presumption was that all producers would meet the standard and, effectively, self-regulate. According to Heckman, during this period the FDA shifted away from demanding analytical data on migration and sought to simplify the process by requiring that petitions include only a description of testing methods used to determine migration.[99]

Heckman assured the Food Packaging Materials Committee that the FDA "recognizes the fact that the Food Additives Amendment of 1958 was not only *not* drafted with incidental additives [indirect food additives such as plastic monomers that migrate into food] in mind, but . . . is . . . inept as a statutory device to regulate food packaging materials." This was because the "analytical methods to determine the amount of incidental additive migration into food are, as a practical matter, impossible to develop." In other words, according to Heckman and some officials at the FDA, plastic packaging was effectively exempted from regulatory oversight by prior use or because of the assumption that chemicals that migrated from plastics into food did so at levels either below or barely above detectable levels and thus presented an insignificant risk to public health. What you can't see won't hurt you. "Whatever the lawyers may think of this evolution," Heckman concluded, "and the degree to which the treatment afforded packaging materials complies with the statutory mandate, and regardless of how the scientific world will accept the new FDA attitude, we are sure that the salesmen among you will be pleased." In the end, Heckman reassured plastics producers, "Packaging is not a threat to the public health."[100]

In the decade after the passage of the 1958 Federal Food, Drug and Cosmetics Act, or the Food Additives Act, the chemical, plastics, and food industry trade associations worked diligently to pull the teeth from the Delaney clause by ensuring that it did not provide legal precedent for a per se interpretation of a hazard. In 1962, the FDA passed the DES proviso, which permitted use of the synthetic estrogen, despite evidence of its carcinogenicity, in livestock provided that no detectable residue could be found in human food. A chemical, therefore, would not be banned simply because it was used in food production per se and found to be carcinogenic. There needed to be quantitative evidence of its presence, which meant that the definition of a food additive and its "safe" use would be tied to changes in analytical chemistry. This generated a considerable problem for the producers of DES and other suspected carcinogens, however, because as methods changed and detection thresholds fell, a chemical's safety theoretically could come into question and the chemical could be subject to a ban.

Industry ideally sought to harmonize the process of setting safety standards for carcinogens and noncarcinogens in a way that upheld the assumption that, at some small concentration, all risks, as Kehoe had remarked in the early 1950s, were manageable and acceptable. Such an approach would stabilize the process of defining safety. Lost in this framework of risk was any distinction between carcinogenicity and toxicity as articulated by men such as Smith and Hueper, who argued that a carcinogen can be low in toxicity but of significant concern when exposure occurs over a lifetime and in the very young.

When public interest groups sought to force the FDA to actively use the Delaney clause to pull carcinogens from the market in the 1970s, they relied on evidence of carcinogenic effects from high-dose studies on adult animals. Their goal at the time was to enforce a per se interpretation of the Delaney clause and ban a few high-profile carcinogens from the food supply. This resulted in a distorted notion that public interest groups sought to ban anything that caused even a single rat to die of cancer. Similar contentions that public fears about carcinogens were wildly disproportionate to their true risks pointed to the fact that in order to replicate animal studies, humans would need to consume large amounts of chemicals, when in fact human exposures were very low. Arguments for keeping carcinogens out of the food supply came to be viewed as anathema to progress. Exposure to pesticides, plastics, and preservatives was described as negligible and as a necessary side effect of modern society. Consideration of the risks of chronic low-level exposures in the young, and thus the added variable of time, subsequently found no place in debates over interpretation of the Delaney clause or the definition of safety. The normative assumptions that "life is full of risks" and that, if risk is managed well, a "little bit can't hurt" would form the logical framework used to evaluate and define chemical safety for decades to come.

2 The "Toxicity Crisis" of the 1960s and 1970s

In 1972, a study conducted by the National Health and Lung Institute found traces of a class of chemicals called phthalates in the blood of its laboratory workers. Phthalates are compounds added to plastics, notably PVC. "We're all a little plastic," declared a *Washington Post* report. Concentrations of 10 to 30 parts per million (ppm) in blood samples might seem inconsequential, but, as the reporter explained, "It is far from a tiny number for a biochemist."[1] We had become what we made, and awareness of this reality raised questions and fears about what these chemicals might be doing to the human body.

Reports such as the one in the *Washington Post* were hardly new. Since the 1950s, hazardous chemicals had been detected throughout the environment and in the human body: strontium-90 in human and cow milk due to atmospheric testing of atomic weapons; the synthetic estrogen DES in meat; and residues of the pesticide DDT in wildlife, milk, and food. The passage of the Food, Drug and Cosmetics Act in 1958, eight years after Representative James Delaney's first hearing on chemicals in foods, coincided with community-led campaigns to block government pesticide spraying in suburban neighborhoods, such as those in Long Island that first caught the attention of Rachel Carson. Over the course of the 1960s, growing public awareness and concern about chemical pollution galvanized an emergent environmental and consumer health movement.

When Carson began her research on pesticides, in the late 1950s, she was a highly acclaimed and nationally recognized science writer with several best sellers under her belt. In her much-anticipated book *Silent Spring*, published in 1962, Carson wove the science of pesticides into a cautionary narrative that warned of humans' potential to self-destruct through heedless and indiscriminate use of poisons. Within months of

43

the publication of *Silent Spring*, two national events brought into stark detail the risks and dangers of modern technology: the near approval of the drug thalidomide and the Cuban Missile Crisis. That summer, FDA medical officer Frances Kelsey became a national heroine for her refusal to accept industry claims for the safety of thalidomide, an antinausea drug prescribed to pregnant women, already distributed in the United States but widely used in Europe. Her wisdom in heeding the evidence of profound birth defects caused by thalidomide use in Europe was spun into a political narrative to promote long-sought drug reforms. The story of thalidomide, and Kelsey's role in blocking its approval, was intentionally leaked by Senator Estes Kefauver's (D-TN) antitrust committee, which for years had been pushing for drug reforms. The thalidomide story, like the disaster of elixir sulfanilamide (see chapter 1), provided a powerful example of the critical problems of the status quo and thus the need for reform. In the case of thalidomide, the drug reform came via the Kefauver-Harris Amendment of 1962, which required that drugs be proven safe *and* effective.[2] Beyond its use as an illustration of policy tragedy, the thalidomide story and its jarring images of deformed infants fueled a deepening public fear of the real dangers of human innovation and ingenuity.

That fall, just months after *Silent Spring* had come into bookstores and reached the best-seller list, Carson was invited to speak at a Kennedy Seminar in the home of the secretary of the Interior, Stewart Udall. The event occurred just days before the Cuban Missile Crisis, when America came terrifyingly close to nuclear war with the Soviet Union.[3] As Linda Lear writes in Carson's biography, "The crisis over the misuse of pesticides was, for her, perfectly analogous to the threat from radioactive fallout and justified her social and political criticisms of the government and the scientific establishment as well as her implicit calm and reasonable call for citizen action."[4]

The publication of *Silent Spring* marked the beginning of a self-conscious modern environmental movement, which coalesced around declension narratives of environmental pollution and threats to democracy posed by what Dwight Eisenhower termed, in his 1961 farewell speech, the "military-industrial complex" and the "scientific-technological elite."[5] Carson, who came under attack by the chemical industry, urged individuals to question who speaks about pesticides and why.[6] The dangers of human innovation lie not just in technology itself but in the failure to check the power of producers of such technology. Viewed from this perspective, the case of thalidomide demonstrates just how close the country had come to widespread use of a dangerous drug: one brave and stubborn woman stood up to industry pressure and thwarted the status quo in regulatory decision making. Over the course of one of the most politically tumultuous decades in American history, popular faith in the wonders of unchecked medical and technological innovation declined and skepticism rose, and subsequently regulatory decision making was brought under greater scrutiny. Public interest groups, including consumer health and environmental organizations, emerged to directly challenge industry and government practices and policies.

For the regulatory agencies, such as the FDA, the 1960s and early 1970s were a period of major expansion, transition, and crisis. As Barbara Troetel argues in her dissertation on the FDA, by the late 1960s and early 1970s the agency's actions had come under considerable public scrutiny because of a combination of consumer advocacy pressure, notably from Ralph Nader's Public Citizen, media attention, new disclosure policies enacted under the Freedom of Information Act (1966), its amendment in 1971, and the Federal Advisory Committee Act (1972), which mandated public participation in federal committees.[7] Public scrutiny of, engagement in, and visibility of regulatory decision making generated a crisis of confidence in the reputation of the FDA. The iconic government heroine, as embodied by Frances Kelsey at the beginning of the decade, would become unimaginable by the early 1970s, when FDA officials, including the commissioner, were losing their jobs as a result of public crises over the safety of chemicals.

The FDA found itself in the public spotlight as it struggled to address the problem of toxic chemicals in the food supply and environment. Increased attention by consumer advocates and environmentalists to these problems were what the chemical industry referred to as the "toxicity crisis" of the 1960s and 1970s. A critical component of this crisis involved determining how the FDA, and later the EPA and the Occupational Safety and Health Administration (OSHA; established in 1970), would manage chemical carcinogens. Are carcinogens hazards per se, or can the risks be managed at low levels—that is, regulated according to a *de minimus* standard? How the FDA interpreted the Delaney clause for chemical carcinogens not only influenced the markets for high-profile chemicals such as DES, the sugar substitute cyclamate, and plastics, as described in this chapter, but also informed how carcinogenic risk would be identified and managed by EPA and OSHA. At the broadest level, agency leaders sought to navigate how inevitable chemical risks, including carcinogens, would be assessed and managed as "safe."

Although there were moments in the early 1970s when strong regulations on carcinogens were issued, these proved to be policy exceptions.

By the end of the 1970s, persistent inflation and rising unemployment contributed to the waning political power of public interest and environmental organizations and the growing acceptance of deregulatory policies. Scientific and regulatory arguments for zero tolerance of exposure to carcinogens yielded to new assessments of carcinogenic risks that would pave the way for setting safety standards, effectively blocking the door to complete product bans. The new prevailing political and scientific assumptions held that all risky chemicals, including carcinogens, could be effectively managed as safe, and regulatory decisions were subject to risk-assessment and cost-benefit analysis. The human body continued to bear witness to the petrochemical revolution because the logic of chemical management remained grounded in the assumption that although we might be exposed to low levels of chemicals, this risk was controllable, manageable, acceptable, and, above all else, safe in the case of the vast majority of chemicals.

DEFINING CARCINOGENS UNDER THE DELANEY CLAUSE

Interpreting the Delaney clause on carcinogens, included in the 1958 Federal Food, Drug and Cosmetics Act (Food Additives Act) and later the Color Additive Amendments of 1960 and Animal Drug Amendment of 1968, became a critical litmus test of the FDA's authority and power. The question raised by the Delaney clause was whether carcinogens posed an unacceptable risk per se, regardless of exposure level, or whether risk was necessarily tied to exposure. A strict interpretation of the Delaney clause would challenge the risk logic of the *de minimus* standard and open the door to numerous product and chemical bans.

The political and scientific disputes over carcinogens and the interpretation of the Delaney clause discussed in this chapter manifested in questions about the relationship between dose and effect. Historian Robert Proctor's seminal book on cancer and environmental exposures, *Cancer Wars*, details the political contours of these debates that divided carcinogens from noncarcinogens.[8] For noncarcinogens, the prevailing presumption about the relationship between the dose and the toxic response held that the effect increased linearly with dose but that at some small dose the effect diminished. In other words, the effect reached a threshold. Graphically, the relationship looked like a hockey stick, with effects leveling off at a certain low level of exposure. The advantage of this model, from the perspective of the regulated industry, which had long supported use of this dose-response relationship in assessing chemical risks, was that even if technological ability to detect chemicals dropped, as it had done for decades, a safety standard could remain constant.

Carcinogens, in contrast, were presumed to follow a completely linear dose-response relationship, so that for all concentrations of the chemical above zero a carcinogenic response would be expected. The scientific and political legitimacy of this assumption became evident in debates over interpretation of the Delaney clause in the early 1960s and throughout the 1970s. Indeed, the inclusion of a distinct clause for carcinogens in the 1958 food law, followed by growing public concerns about cancer in the 1960s and 1970s, continued to provide support for distinct characterizations of the dose-response relationship for carcinogens. The question inherent in this model, which manifested in debates over implementation of the Delaney clause, was this: How do you define zero? Defining zero in absolute terms, as some proponents of the Delaney clause sought to assert, would mean that a carcinogen was a carcinogen regardless of evidence of exposure—it was a carcinogen per se. Applying this framework to interpretation of the Delaney clause threatened to significantly broaden the regulatory reach of the FDA because any chemical carcinogen in food production, processes, or transport potentially could be restricted from the food supply. The implication for chemical producers was obvious: a *de minimus* standard approach allowed the continued use of chemical carcinogens in food provided that the exposure was undetectable.

The FDA wrestled with these opposing interpretations of the Delaney clause and sought to clarify its policy on carcinogens in a number of specific cases, notably those of DES, cyclamates, and several plastics. Understanding how and why these compounds were kept on the market in the 1960s but later withdrawn reveals a consistency in policy making under the *de minimus* standard even in the face of increased public pressure and public interest scrutiny in the 1970s.

THE CASE OF DES

By the early 1960s, Jerome Heckman, lead attorney for the Society of the Plastics Industry (SPI), had confirmed that the FDA had no intention of using the Delaney clause to ban all chemical carcinogens.[9] The agency would regulate safety according to the *de minimus* standard: in other words, zero would not be interpreted in absolute terms (a chemical would not be a carcinogen per se) but would be tied to detectable levels and evidence of significant risk. This interpretation became official policy as the agency determined how to regulate the use of DES as a feed additive in livestock. This policy approach resulted in the approval of, as well as the eventual banning of, DES as a food additive.

In her history of the FDA's decisions on DES, *Toxic Bodies*, Nancy Langston details the agency's changing position on the drug's safety from the 1940s until its ban in the 1970s, arguing that during this period the agency moved away from its original precautionary approach to the drug's risk, which, if heeded, would have better protected public health. Langston points to a number of potential reasons that the agency came to accept the risks of DES, including gender biases, industry and political pressure on the agency given the size of the DES market, and industry "capture" of the regulatory leadership.[10] While recognizing that all of these factors came into play in the agency's decisions on DES's use, particularly as a human pharmaceutical, this section focuses more specifically on how the agency's interpretation of the Delaney clause provided a legal framework for accepting the perceived risk of DES as an animal feed additive. In the case of DES in animal feed, the agency's interpretation of the Delaney clause as a *de minimus* standard was relatively consistent from the early 1960s up to the chemical's eventual ban in the late 1970s. What ultimately resulted in the removal of DES from the market was a combination of increased scientific scrutiny, public pressure, and applied use of advances in analytical chemistry, rather than a shift in the legal interpretation of the *de minimus* standard.

In its first step to regulate DES under the new Food, Drug and Cosmetics Act of 1958, the FDA provided manufacturers with a "prior use" exemption—a grandfather clause in the law. This meant that, in the case of DES, existing manufacturers of the drug could continue production but any new applications were barred. The rapid approval of DES by the regulatory agency, according to letters in 1960 from Mel DeMunn of Farm Report, U.S.A., to Representative James Delaney, the political champion of the contentious cancer clause, and Representative John Dingell (D-MI), "came through strong pressures from the top. That stilbestrol clearance was ordered from the top is no secret."[11]

However, the prior use exemption resulted in what legal counsel from Pfizer called "inequitable" competition, since it excluded use by other manufacturers. (Edward Charles Dodds, who first identified DES, never patented it.)[12] To remove this market barrier, drug and chemical industry representatives successfully lobbied for a proviso to the Federal Food, Drug and Cosmetic Act in 1962 that allowed expanded use of DES in livestock by other companies. This marked a significant first step toward completely removing the Delaney clause, which was a priority of the

Manufacturing Chemists' Association (MCA. In a letter sent to Dingell and carbon-copied to Delaney in 1960, one Edward Harris (whose affiliation is not known) wrote, "The Manufacturing Chemists Association wants it [the Delaney clause] out—period. Both associations [the MCA and Abbott Laboratories] have called for a fresh study by a committee of scientists. . . . It seems logical to assume that the industries are calling for a new study of carcinogens because they feel they could salt the committee with scientists who front for them, and get a report quite different from the Rome recommendations [the 1956 International Union against Cancer statement that called for all nations to restrict carcinogens from the food supply]."[13]

While falling short of removing the Delaney clause, the DES proviso permitted the use of the drug in livestock, provided no detectable residues were present in edible portions of meat. The reasoning: while DES might be carcinogenic, if it was not detectable in tissue, no exposure could be assumed, and thus the drug could not be subject to regulation. The detectable limit was set at 2 parts per billion (ppb), using the mouse uterine assay, the standard test for identifying estrogenic activity. As such, the DES proviso interpreted the Delaney clause as applicable only to quantifiable amounts of carcinogens. If you can't detect it, you can't regulate it. Residues of DES that fell below the detection limit were presumed to be safe. This interpretation allowed for the continued use of DES, even in the face of scientific evidence of its carcinogenicity in laboratory animals in the 1960s. Applying this interpretation of the clause, the FDA restricted the use of DES in poultry after the U.S. Department of Agriculture detected residue in the edible tissue of chickens but allowed continued use in cattle and sheep, where residues had yet to be identified.[14]

The DES proviso represented a significant step toward harmonizing the Delaney clause with the *de minimus* standard for all other noncarcinogenic food additives. At the same time, however, it tied the definition of a carcinogen and the legal trigger for a ban to the advancing technology of chemical analysis and detection. For industry trade associations interested in keeping markets open, this was a significant step, but it fell short of their desire to completely eliminate the clause. And by the mid-1960s, calling for the complete elimination of a legal clause that sought to protect the public from cancer-causing chemicals had become a politically impossible position. Drawing their political strength from public concerns about chemical pollution and industry capture of government agencies, environmental and public interest organizations began to directly challenge the FDA's interpretation of the Delaney clause.

CHEMICAL SAFETY IN CRISIS: CHALLENGES
TO THE DELANEY CLAUSE

With the successful passage of the DES proviso and the FDA's informal policy of applying the Delaney clause only to significant and detectable amounts of food additives, there appeared to be momentary consensus on how to define chemical safety. The rising political influence of environmental and public interest organizations effectively disrupted this harmony. Popular books such as Carson's *Silent Spring* and Ralph Nader's *Unsafe at Any Speed* (1965) drew sharp attention to the failure of the government to protect public health and the environment. In *Silent Spring*, Carson writes of the waves of chemical poisonings among birds exposed to DDT. She highlights research that detected the chemical in the testes, egg follicles, and ovaries of robins and in dead birds, and she repeats the concerns expressed by Wilhelm Hueper of the National Cancer Institute (NCI) and the International Union against Cancer in the 1950s. "Experiments," she writes, "show that the younger the animal is when it is subjected to a cancer-producing agent the more certain is the production of cancer."[15] That carcinogens might function differently than toxic compounds and present a particular risk to the young was the scientific argument behind the original call to keep them out of the food supply and subsequently became an important argument for absolutely restricting carcinogens. By the late 1960s and early 1970s, the development of well-organized public interest groups, such as Public Citizen, the Environmental Defense Fund, the Natural Resources Defense Council, and others, provided the means to direct these concerns toward regulatory decision making at the FDA and offered political support to those within government institutions who championed strong policies on carcinogens.

To question the safety and necessity of pesticides and toxic chemicals was to question political and economic decision makers. Carson's disturbing tale of chemical contamination focuses explicitly on the moral predicament of this pollution. Who decided our fate? she questions. "Who has the *right* to decide[?] . . . The decision is that of the authoritarian temporarily entrusted with power."[16] *Silent Spring* marks the beginning of the modern environmental movement because it integrates narratives of uncertainty, skepticism, and fear about the darker underbelly and side effects of human progress: industrial chemicals, the threat of nuclear war, radiation fallout from nuclear weapons testing, and the loss of natural spaces and wildlife, together with growing dissent against mainstream power structures: the military-industry complex.

Dissent and the questioning of authority and existing power structures lay at the heart of many social movements of the 1960s and 1970s, from civil rights, feminist, antiwar, and counterculture movements to environmentalism. These movements sought to restructure society and as such looked to reform, rather than overthrow, the state. For their part, mainstream environmentalists sought to strengthen state authority over markets as part of the solution to pollution. They worked toward the expansion of environmental laws and the establishment of new agencies. Subsequent struggles and debates over whether to restrict chemicals from the market reflected larger debates over the appropriate role of the state in protecting public health and intervening in the private sector.

The establishment of public interest organizations brought increased public scrutiny to regulatory decision making on the safety of industrial chemicals. Harold Stewart, a colleague of Wilhelm Hueper's from the NCI, noted the changing political climate within the agencies in a letter to Hueper in 1973. "Things are improving somewhat. . . . Nader has made all the difference in the world. No longer is industry or the government organizations that deal with industry and formerly favored industry able to conceal nefarious activities today, at least not to the extent they might have twenty and thirty years ago."[17] With public pressure from consumer advocates such as Nader, dissent within the agencies became possible. In particular, researchers within the FDA spoke publicly about internal practices, data manipulation, and perceived failures to adequately respond to public health threats. Critical questioning of FDA practice and policy by agency staff led to the first chemical ban (that of cyclamates) under the Delaney clause and increased pressure to ban DES.

The key FDA whistleblower was Howard Richardson, a pathologist and chief of research at the FDA in the 1950s and 1960s. For years, he had worked in vain to improve the research and resources of the agency's pathology department. (Improving the agency's pathology department presumably would have strengthened the agency's accuracy and advancement in disease detection and diagnosis.)[18] Richardson outlined his concerns about scientific practices at the FDA in a memo to the agency's commissioner, Herbert Ley, in 1969. In his memo, Richardson alleged that throughout the 1950s, during George Larrick's long tenure (1954–66), the agency had manipulated laboratory tests on cyclamate, even overlooking evidence of its carcinogenicity, so as to sustain its approval for use as a food additive.[19] Ley had become commissioner in 1968, replacing James Goddard, who had served for two years after Larrick's departure. Richardson described a disgraceful scientific climate at the FDA, listing further

examples, including the agency's failure to complete studies on monoso-
dium glutamate (MSG) and its refusal to publish its own study raising
questions concerning the safety of the pesticide heptachlor.[20]

Richardson's memo was reported by the press and circulated to members
of Congress, including Representative Delaney. Public attention focused
on the potential carcinogenicity of cyclamates and the FDA's regulation
of these food additives as "generally recognized as safe" (GRAS) substances under the 1958 law. Reviews of cyclamate research by the NCI,
the National Academy of Sciences (NAS), and the FDA cited evidence
of bladder cancer in laboratory animals exposed to very high doses of
the chemical. Ley responded by releasing a public statement cautioning
the public to limit their cyclamate intake.[21] One of the FDA researchers,
Jacqueline Verrett, who was involved in a cyclamate study that found
deformities in exposed chick embryos, appeared on the *NBC Nightly News*
and advised pregnant women to avoid the sweetener altogether, likening
the situation to the thalidomide crisis.[22] The agency's message to the public
was profoundly confusing. Richardson's disclosure of the agency's failures
struck a serious blow to its integrity and credibility. Ley's warning and
Verrett's strong indictment of cyclamate were a reversal of the FDA's long-
held position that the substance was safe, which lent further legitimacy to
claims that the agency had failed to protect the public. The agency's about-
face decision on cyclamate safety profoundly undermined the reputation
of the agency. Facing consistent public pressure, evidence of cyclamates'
carcinogenic effect, and recommendations from the NCI to remove the
product from the market, the secretary of the Department of Health, Education and Welfare (HEW), Robert Finch, stepped in to resolve the crisis,
announcing a partial ban on the sweetener in 1969.[23]

For the first time since the passage of the Delaney clause, the FDA
had banned a highly lucrative additive from the market despite its GRAS
listing. The industry conceded and chose not to fight the decision in court.
(This was not the first or last time the agency banned sweeteners. In 1950,
before the Delaney clause, the FDA had banned two sweeteners found to be
carcinogenic, and it issued this decision under the general safety standard
of the 1938 law.) But this was the last time that industry would choose not
to fight the ban legally and politically. In the wake of the cyclamate ban,
the market for the artificial sweetener saccharin (also a GRAS substance)
boomed, and in 1977, the FDA initiated a politically charged process to
consider a ban in a remarkably different political climate.[24]

But in 1969, the FDA's cyclamate ban set a very low mark for the
agency. As the magazine *Science News* reported at the time: "The agen-

cy's on-again, off-again attitude toward cyclamates also cast doubt on its operations and caused considerable embarrassment to Health, Education and Welfare Secretary Robert H. Finch." Indeed, the reporter noted, "It is difficult to find anyone who is happy with the FDA."[25] For consumer advocacy and environmental groups as well as the public at large, the agency appeared to be failing to protect people from hazardous chemicals. Cyclamates went from being considered GRAS substances to chemical carcinogens. From the perspective of the regulated industry, the FDA was making decisions about chemical safety based on political pressure. The agency was caught in a crisis of legitimacy. Public disclosure of suppressed scientific research fueled environmental activists and consumer concerns about regulatory capture, while efforts to respond to these concerns angered the industry trade associations, which saw the agency as bending under public criticism and perpetuating a decline in consumer confidence in the safety of chemicals in food.

Ley had inherited an agency in major transition. The Kefauver-Harris drug reform law of 1962 and, to a lesser extent, the Food Additives Act had created considerable new responsibilities for the agency in a very short period. These regulatory responsibilities expanded at a time when public scrutiny of the decision-making process was increasing. In yet a further blow to the agency's competency, President Nixon directed the FDA to review all of its GRAS substances. For Finch, the cyclamate crisis was a critical failure of leadership, and he fired Ley. Howard Richardson, the FDA pathologist and whistleblower, was afforded no protection and was demoted from his thirty-year appointment as chief of the research division, despite confirmation of his allegations by another FDA patholo-gist, Kent Davis.[26] To salvage the agency's reputation, Finch hired Charles C. Edwards as the new commissioner. Congressman Delaney accused the FDA and HEW of punishing Richardson for his criticism of the agency and strongly objected to Edwards's appointment.[27]

Finch hired Edwards for his strong managerial skills and solid rela-tions with industry. Prior to accepting the position at the agency, Edwards had worked for the private consulting firm Booz Allen Hamilton.[28] In the mid-1960s, the MCA had contracted with that firm to develop an orga-nizational strategy for creating a new institute on environmental health and to manage, distribute, and develop environmental health research.[29] As Congress made new commitments to environmental health research with the establishment of the National Institute of Environmental Health Sciences in 1964, the MCA worked to position itself as a source of scientific authority and expertise. Trained as a medical doctor, Edwards served as

FDA commissioner from 1969 to 1973, when he was appointed assistant secretary of health at HEW.

Edwards directly confronted the need to better manage the FDA's relations with consumer advocates and to improve the transparency of the agency's decision-making process. Recognizing the real threat of consumer advocates, who had deep skepticism and concern about his leadership because of his ties to industry, he established a consumer advisory group and began to call on outside experts. With the support of his general counsel, Peter Hutt, the agency responded to new demands made under the Freedom of Information Act. Under Edwards and Hutt, the agency's message was that decisions would be based on rational evaluation of the data and not on emotionally charged public responses to disclosure of risks—the perceived lesson of the cyclamate ban. Edwards was keenly aware of the challenge the agency faced from increased public engagement in the regulatory process and rapid changes in scientific research and technologies. "As the biological and physical works probe deeper and learn to measure values ever more minutely, our common concepts of what is . . . safe must change."[30]

The Delaney clause, according to Edwards, was an emotionally driven response to risk because it demanded absolute safety. In particular, he argued, it was seriously flawed because it forced the agency to adhere to the unscientific "single rat" argument: strict interpretation of the clause meant that if just one rat developed cancer, even a supposedly beneficial chemical could be banned.[31] Absent from this simplified articulation of the clause were Hueper's concerns about the particular vulnerability of the young to chronic exposure and the ability of some chemicals to be carcinogenic at nontoxic levels.

Edwards was not alone in considering the Delaney clause to be highly unscientific. In addition to the MCA, the FDA's director of toxicology, Leo Friedman, and Peter Hutt shared this view. Hutt had come to the FDA in 1971 from the law firm Covington and Burling, whose senior partner was a member of the Food Law Institute, which represented the interests of the food and chemical industries. Keenly aware of Hutt's position on the cancer clause, Delaney vehemently protested his appointment to Edwards, citing the law firm's sympathies with industry interests and opposition to the Delaney clause.[32] Delaney's charge of conflict of interest probably raised little concern for the commissioner, who arguably found the congressman's complaints and his clause a distraction to the agency's work.

In a speech at Temple University in Philadelphia in May 1970, Commissioner Edwards suggested that the problems faced by the FDA could be solved by getting "the Ralph Naders, our Congressional critics, and even some of the 'experts' within Government structure to leave us alone" so that the agency could make necessary reforms, among them changes to the Delaney clause. Such changes, Edwards remarked, would "permit some scientific rationality in making these decisions. If we were to apply the criteria of the Delaney Amendment *[sic]* across-the-board eventually we would be reduced to a nation of vegetarians and even some of the vegetables would have to be banned."[33] Edwards favored an approach that would allow the agency to set safety standards for carcinogens, and under his leadership the FDA "hinted at the possibility of a threshold for saccharin's carcinogenic effect."[34]

Edwards's derisive mention of the "Naders," "Congressional critics," and "experts" in government who were stirring up trouble for the agency was a reference to consumer advocates, Congressman Delaney, who had called for an investigation of FDA, and, most likely, Umberto Saffiotti at the NCI.[35] Saffiotti had been a member of the 1956 Rome meeting of the International Union against Cancer that called for all nations to restrict carcinogens from the food supply. In 1968, he joined the NCI in the midst of the cyclamate controversy, serving there as an associate scientific director for carcinogenesis and etiology in the Division of Cancer Causation and Prevention. It was Saffiotti who, after reviewing the cyclamate data, recommended a ban to Secretary Finch. After the controversial ban was issued, Finch appointed Saffiotti to chair a scientific investigation of the Delaney clause. Saffiotti's committee, the Ad Hoc Committee on the Evaluation of Low Levels of Environmental Chemical Carcinogens, reaffirmed the scientific legitimacy of the Delaney clause and concluded that "no level of exposure to a chemical carcinogen should be considered toxicologically insignificant for man" (i.e., no threshold of safety existed for carcinogens). The final report dismissed charges of scientific illegitimacy, or the "single rat" argument, leveled at the clause.[36] The Delaney clause, the committee concluded, "allows the exercise of all the judgment that can safely be exercised on the basis of our present knowledge."[37]

The findings of Saffiotti's committee were countered by a position paper, "Guidelines for Estimating Toxicologically Insignificant Levels of Chemicals in Food," released that same year by the Food Protection Committee of the NAS. Its focus was to present risk-benefit analysis as an alternative approach to a zero-tolerance policy for carcinogens. According to Hueper and researchers at Public Citizen's Health Research Group, the Food Protection Committee disproportionately represented the interests of food and chemical companies. The committee, chaired by William

Darby, an outspoken critic of Carson's *Silent Spring* and president of the Nutrition Foundation (an organization allegedly financed by the food and chemical industry), resoundingly denounced the Delaney clause and the nonthreshold model for carcinogens.[38]

The NAS committee concluded that threshold effects existed for all chemicals, including carcinogens. This meant that exposure to carcinogens at "insignificant" levels of exposure could be safe—a conclusion at odds with the Saffiotti report and a strict interpretation of the Delaney clause. By contending that the dose-response relationship for all chemicals followed a threshold model, the NAS report lent legitimacy to a policy of establishing safety standards for carcinogens. The report went a step further by recommending that chemicals in production for more than five years without any indication of harm could be considered safe in small amounts, a recommendation that blatantly ignored evidence of the long latency of cancer. Approval of the report came from the highest levels of the NAS, including the president of the academy, Philip Handler, who publicly criticized Saffiotti's committee as upholding an unscientific policy.[39]

Despite rebuke from the prestigious NAS, Saffiotti refused to be reticent. In 1970, together with other scientists at the NCI, he submitted research on the carcinogenicity of DES to the surgeon general.[40] The NCI had long expressed concerns about the drug and its use in animal feed. In the 1950s, as head of the Environmental Cancer Section, Hueper had warned of the carcinogenicity of estrogens; in 1960, Roy Hertz, the chief of endocrinology at the NCI, had reported that DES "is known to produce a variety of tumors in several species after prolonged exposure"; and in 1963 researchers at the NCI documented tumors in newborn mice exposed to a single injection of DES. However, all the data in these studies came from experimental studies on laboratory animals.

The tenor of the debate changed significantly in 1971, when the first study of the effects of DES on humans was published. A paper by researchers at Massachusetts General Hospital and Harvard University reported an increase in rates of a rare vaginal cancer in young women exposed to DES (taken as a prescribed drug) while in their mothers' wombs.[41] Not only had the study found significant evidence of DES's carcinogenicity, the effects had occurred at exposure levels considered therapeutic and safe. Moreover, the effects appeared decades after exposure. The tumors reported were quite rare and the association with DES exposure was strong. Laboratory research together with epidemiological findings had now provided considerable evidence for DES's adverse effects in humans. The findings also lent greater scientific legitimacy to the very concerns Hueper had articulated before Congress in the 1950s: the very young were most sensitive to the carcinogenic effects of chemicals, even at very low, nontoxic levels.

Epidemiological evidence of carcinogenicity in the daughters of women who had taken DES resulted in an immediate warning from the FDA regarding the use of the drug during pregnancy. For the next several years, additional data confirming carcinogenic and reproductive risks to exposed daughters, as well as data on reproductive risks to exposed sons and increased breast cancer risk to women who took DES, led to greater restrictions on its use.[42] As for its use in cattle, however, Edwards attempted to seek ways to keep DES on the market—"safely and effectively."[43] DES's carcinogenicity in humans did not necessarily trigger a ban on the drug's use in food production. The FDA had long permitted DES as a feed additive provided it was undetectable in edible tissue, and Edwards sought to continue this use. Detection, therefore, was key to determining whether the FDA would ban DES in food production.

Therefore, when DES was detected in the livers of cattle, the agency was forced to issue a limited ban on its use in livestock feed in 1972. The ruling affected DES only in feed use not in its implantation into animals. According to Edwards, the inappropriate language of the Delaney clause forced the agency to take action to withdraw any new applications of DES in animal drug use. From his perspective, he acted, not because of evidence of a risk to public health, but rather because the letter of the law left him no choice.[44] Two DES manufacturers challenged the ruling in court, and in 1974 the U.S. Court of Appeals decided that the rule was invalid because the agency had failed to hold a public hearing on the issue. The ban was subsequently lifted. However, detectable levels of DES continued to appear in animal tissue even with the use of implantation.[45]

DES was detected in edible tissue because people were looking for it and because of a change in technology. The method of detection used by the FDA was the mouse uterine assay (i.e., test), which identified the presence of any estrogenic compound, not only DES, at levels as low as 2 ppb. Manufacturers contended that an increase in exposure to an estrogen could result in a carcinogenic effect but also that a low level existed at which estrogenicity might be present but no carcinogenic response would result. In other words, estrogens were presumed to follow a threshold dose-response curve. Officials at the U.S. Department of Agriculture began using gas liquid chromatography/mass spectrometry, which can detect DES at levels below 2 ppb, effectively challenging the use of the uterine assay and triggering a ban. Political pressure and new evidence of detection, therefore, forced the hand of the agency. After more than a decade of debate, the agency issued a total ban on DES use in animals in 1979.[46] Looking back on the decision many years later, Donald Kennedy, who served as commissioner of the FDA at the time, argued that the agency could have reached the same decision on DES without the Delaney clause by using the general safety standard of "reasonable certainty of no harm" outlined in the Food Additives Act.[47]

Early in the 1970s, in the midst of legal and political debates over the interpretation of the Delaney clause, members of Congress, health and environment advocates, the FDA, and chemical and food industry representatives made numerous attempts to clarify or reform the law. Beginning in 1969 and continuing into the mid-1970s, the liberal Democratic senator from Wisconsin Gaylord Nelson, father of the first Earth Day in 1970 and a supporter of the inclusion of the Delaney clause in 1958, introduced a number of bills to expand and strengthen the clause. Nelson's legislation sought to extend the scope of the Delaney clause to include health effects beyond cancer, such as reproductive, developmental, and immunological toxicity, and to expand the FDA's power and authority by requiring industry to define chemical benefits and allowing the agency to inspect factories. Counterefforts to eliminate the clause included bills introduced in the House by Representative William Scherle (R-IA) that would have allowed the FDA to establish safety standards for carcinogens.[48] In addition to legislative activity, ongoing congressional hearings, led by Senator Edward Kennedy (D-MA), Senator Nelson, and Representative John Moss (D-CA), as well as investigations by the General Accounting Office of the FDA's approval of food additives, such as saccharin and the color additive Red No. 2, focused considerable political attention on the FDA and its regulatory decision-making process for chemical carcinogens.[49]

As exemplified by the agency's decision-making process on DES in livestock, improvements in detection technologies presented a considerable challenge to establishing stable safety standards for carcinogens. The legislative push to allow the agency to set such standards attempted to resolve this predicament. Without any real opportunity for legislative reform, the agency tried to issue an internal policy to define "no detection" or "no residue" according to a small lifetime risk of cancer, thereby ending the problem presented by falling detection limits. The proposed policy, the "Sensitivity of Method" (SOM) proposal, basically would have allowed the establishment of safety standards for carcinogens based on some marginal level of risk and would have halted "the endless chase of analytical 'zeroes.'"[50] In the 1970s, however, proposing that some cancer risk existed and yet was not absolutely avoidable was not politically palatable. The SOM proposal sat in mothballs for fifteen years until the FDA issued its Regulation of Carcinogenic Compounds Used in Food-Producing Animals, which promulgated the SOM approach to defining "no residue." As the FDA declared in 2002, when it revised its definition of "no residue," the SOM approach provided steps by which a producer of a carcinogen could obtain FDA approval to use it as a feed additive. In short, under the current policy if "no residue" is detected in the edible portions of tissue by methods approved by the secretary of the Department of Health and Human Services, the additive is cleared for use.[51] Under the revised no-residue definition, a lifetime cancer risk of one in a million is defined as "no significant risk"; therefore, if the concentration of a carcinogen falls below the level at which the risk is one in a million, the additive is deemed safe. It is the producers' responsibility to submit to the FDA the regulatory method used to detect the chemical.[52]

In the 1970s, at the height of public concerns about cancer and toxic substances, arguing for an acceptable level of cancer risk did not, as noted, sit well with the general public or some members of Congress. Further, such a position put one at direct odds with environmental and consumer organizations, which continued to wield considerable political power. Though the FDA's 1972 proposal fell short of becoming official policy, the agency did begin to integrate quantitative risk-assessment methods to evaluate cancer risks—an approach that would dominate chemical regulatory practice by the 1980s and would become institutionalized by the 1990s.

THE CASE OF PLASTICS

"At best," a representative of the plastics industry predicted in 1973, the toxicity crisis of the 1970s "could prove to be just another nagging pain in the neck to an industry already beset by image-damaging problems. At worst, it could limit or halt the projected growth of plastics in some promising end-use markets."[53] The crisis was fueled by political pressure from a well-organized public interest and environmental community, growing public fear and awareness of chemical carcinogens, and expanding regulatory authority over chemicals.

The 1970s was the decade of modern environmentalism. In his first act of the decade, President Nixon signed into law the National Environmental Policy Act, which called for the formation of the Council on Environmental Quality. On April 22, 1970, twenty million Americans celebrated the first Earth Day, and later that year the EPA and OSHA were established. Nixon also declared "war" on cancer and in 1971 signed into law the National

Cancer Act, which established a national testing program for carcinogens at the NCI. Popular awareness of the existence of thousands of chemicals in production, coupled with little to no understanding of the hazards of exposure to them, brought more and more compounds under public scrutiny, from PCBs to PVC. J. Clarence "Terry" Davies, a young political scientist working at the Council on Environmental Quality, began crafting legislation for a comprehensive EPA program to manage industrial chemicals. In 1976, after five long years of debate and negotiation, President Ford signed into law the Toxic Substances Control Act, which gave the EPA the authority to regulate industrial chemicals.

With popular attention focused on the health of the environment and the hazards of chemical production, the chemical industry faced serious challenges to some of its largest markets. The Delaney clause had been used to ban chemicals, effectively opening the door for future market restrictions. Public interest groups, including Public Citizen's Health Research Group, expanded the scope of their campaigns, pressing the FDA to reconsider the safety of chemicals used in plastic packaging, including PVC; phthalates; acrylonitrile, a plastic used in soda bottles; and other, unknown compounds that might, for example, be migrating from new plastic cooking bags.[54]

For environmental advocates, plastics epitomized everything that was wrong with the nation's petroleum dependence. Plastics were charged with polluting the landscape, filling landfills, and driving America's insatiable demand for petroleum and petrochemicals at a time when the country was becoming increasingly dependent upon foreign sources of oil. Communities across the country, including New York City, initiated legislation to ban or limit nonreturnable plastic bottles, which recently had come on to the consumer market. In his 1971 book *The Closing Circle,* Barry Commoner—biologist, environmental advocate, and founder of Center for the Biology of Natural Systems—characterized plastics as examples of how "modern industrial technology has encased economic goods of no significantly increased human value in increasingly larger amounts of environmentally harmful wrapping."[55]

In addition to the nonbiodegradability of plastics, Commoner details rising concerns about the toxicity of PVC and plasticizers that leach from the plastic. Specifically, he briefly highlights the research of Dr. Robert Rubin of Johns Hopkins Hospital, who documents the leaching of plasticizers from polyvinyl blood transfusion equipment. Rubin detected the plasticizer, phthalates, in the blood, urine, and tissue of patients who received blood stored in polyvinyl bags. "Subtle toxicities" of phthalates, such as cell death or increased cell growth, Rubin concludes, are important areas for further research, given "the increasing use of plastics in medical, pharmaceutical, and cosmetics devices over which there currently exist no federal control or regulations."[56]

The plastics trade association, the SPI, was quick to respond and defend the safety and ecological benefits of their material. Working with the public relations giant Hill and Knowlton, the SPI launched a campaign called "Plastics Not Pollution" that redefined plastics in the terms of the ecological discourse of the time. In carefully placed opinion pieces, on television shows, and in the pages of the trade journal *Modern Plastics*, the industry repeated the same message: plastics' contribution to landfills was miniscule, and plastic waste provided a future source of energy—that is, they could be burned in incinerators. Future generations, industry trade association officials argued, could mine the garbage dumps of the past and safely burn plastic—an argument that conspicuously overlooked the known problems of carbon emissions and air pollution (e.g., dioxin produced by the burning of PVC).[57] Indeed, the trade association stated that burning plastic produced only carbon dioxide and water.[58]

But pollution was not the only concern environmentalists raised about the increased production of plastics. Emerging evidence of the carcinogenicity of vinyl chloride, used in the production of PVC, confirmed findings in studies conducted by Hueper in the 1950s. "The growing introduction of various plastics into the human economy, including packaging material of foodstuffs, occupied my interest and efforts for many years," Hueper wrote in his unpublished autobiography in the late 1970s, near the end of his life. "The recent demonstration of liver cancers in polyvinyl chloride workers apparently has furnished confirmatory evidence in support of my interpretations and thereby has demonstrated the urgent need for instituting preventative and protective measures in the plastics industry, and for issuing regulations as to limitations and standards in the use of plastics as wrapping materials of foodstuffs, when direct contact with the food is involved."[59]

PVC proved to be a particularly troublesome problem for the plastics industry. Just as the plastic was making inroads into food packaging with the development of plastic bottles in the early 1970s, the safety of vinyl chloride came under intense scrutiny from regulatory agencies, researchers, and environmental advocacy organizations. As historians Gerald Markowitz and David Rosner meticulously detail in *Deceit and Denial*, mounting evidence of the carcinogenesis of PVC during the early 1970s threatened the massive market for PVC and the FDA's long-held presumption that the plastic was safe for use in food packaging. Alerting the public and the regulatory agencies to the dangers of vinyl chloride presented serious liability issues for the industry and threatened to destroy the booming PVC market. Given the tremendous economic stakes, representatives of the MCA and the SPI allegedly withheld laboratory studies that confirmed the carcinogenicity of vinyl chloride from government officials.[60]

While the MCA scrambled to contain the damaging research on vinyl chloride in 1971 and 1972, the FDA was considering a petition for a new PVC bottle for alcoholic beverages. In 1973, FDA studies on the migration of vinyl chloride from such bottles found detectable levels of the compound leaching from the plastic. On the basis of scientific evidence of the compound's carcinogenicity in high-dose studies, and its migration out of the bottles at detectable levels, the FDA denied the petition for alcohol bottles according to its interpretation of the Delaney clause.[61] The industry's "toxicity" problem was getting worse.

For the next several years, the crisis spiraled nearly out of industry's control: in 1974, newspapers across the country reported that vinyl chloride exposure was responsible for the deaths of four B. F. Goodrich workers from a rare liver cancer, the same cancer found in industry laboratory studies undisclosed to government officials. Given the new and alarming evidence of carcinogenicity, Public Citizen's Health Research Group called for a total ban on PVC in food packaging in 1975, and officials at the NCI released a public statement in support of the proposal.[62] Despite mounting evidence of vinyl chloride's carcinogenic effect in humans and laboratory animals, the NCI's support of a total ban was dismissed by industry officials and some governmental officials as radical and scientifically ungrounded. Herbert Stockinger, the chief toxicologist at the National Institute for Occupational Safety and Health and chairman of the committee on threshold limits of the American Conference of Governmental Industrial Hygienists (established in 1938 to develop threshold limit values for the workplace), called a PVC ban an "irrational decision" and the NCI officials a group of "old fogies."[63]

Heckman, the lead attorney for the SPI, called the Health Research Group's proposal a "cheap shot" that "can do only one thing—cause panic." And panic, he argued, threatened to trigger a decline in consumer confidence that could further weaken a struggling economy.[64] As part of its full-court press to stop the FDA from considering a total ban, the chemical trade association sponsored a study that predicted a PVC ban would result in the loss of 1.6 million jobs and cost the industry millions.[65] Dire estimates of the economic costs associated with environmental controls created a powerful counterargument to expanding regulations at a time of rising unemployment, slowing economic growth, and an energy crisis. Indeed, arguments that regulatory restrictions and bans would further exacerbate rising inflation and undermine a weak economy became a common theme in arguments against stronger environmental and public health regulation.

Ultimately, the chemical industry successfully averted a total ban on PVC in food packaging by demonstrating to the FDA that migration levels of vinyl chloride fell below detection limits. As had been the policy with DES, and consistent with the FDA's interpretation of the Delaney clause, the lack of detection meant that the chemical was not subject to market restriction. The FDA did, however, maintain its restriction on PVC use in alcohol bottles, where vinyl chloride was detectable. Even while the SPI lost a court appeal of the safety standard set by OSHA for vinyl chloride as evidence of its carcinogenicity strengthened, the trade association successfully secured PVC's future in plastic packaging.[66]

REGULATING THE REGULATORY AGENCIES

By the mid- to late 1970s, the popular antipathy toward business that had characterized the political climate of the late 1960s and early 1970s was waning.[67] The long post–World War II economic boom had ended. Rising inflation and unemployment—the conditions of stagflation—brought an end to Keynesian economic policies, which had favored government spending and price controls and had shaped past economic policies such as the New Frontier of John F. Kennedy and the Great Society of Lyndon B. Johnson. As the economic downturn deepened, political support for neoliberal, free-market policies that called for smaller government, lower taxes, deregulation, and the end of price controls strengthened.[68]

While on the one hand Nixon established new regulatory agencies, the EPA and OSHA, on the other hand he put into place a significant check on their authority. In 1971, Nixon established the Presidential Quality of Life Review, which gave the Office of Management and Budget (OMB) the authority to review all regulatory decisions and to evaluate rules' cost-benefit impact. The OMB would thus rein in the regulatory power of the agencies.

Developed by the Army Corps of Engineers in the 1930s to assess flood-control projects, cost-benefit analysis provided a model to assess the economic impact of a proposed project or regulation.[69] Economists, business experts, and legal scholars at the University of Chicago in the late 1950s, representatives of what was referred to as the Chicago School, sought to promote cost-benefit analysis as a means to rationalize regulation and policy decision making. Applying a price to the benefits and costs of a proposed project or regulation was presented as a means to limit the power of the state and to avoid excessive regulation. The shared political philosophy that characterized the Chicago School held that the state should support and protect private property, individualism, and free choice. The market, therefore, should be minimally (if at all) burdened by regulatory oversight. In a cost-benefit analysis, monetary value is assigned to all benefits and all anticipated costs, including lives lost. When the benefits of a decision exceed its costs, the state is said to have acted rationally.[70]

As early as the mid-1960s, the MCA developed public relations campaigns to promote the use of cost-benefit analysis as a more "balanced" approach to regulating air pollution.[71] The newly established EPA adopted practices of cost-benefit analysis to establish clean air and water standards. Since air and water pollution and their associated risks could not be eliminated, regulatory agencies needed cost metrics to assess the balance of costs and benefits to justify action.[72] Importantly, by the mid-1970s, cost-benefit analysis provided an alternative to what was perceived by regulated industry to be an emergent problem plaguing the regulatory process: tort law, in which the courts distinguished a hazardous compound from a safe one.[73] Developing methods for quantitatively differentiating what was safe from what was hazardous within the context of associated benefits and costs provided a process to avoid regulation through litigation—an expensive and timely process.

Determining "benefit" in this equation remained exclusively within the domain of industry. As defined in a 1975 NAS report on evaluating chemicals, "The identification of benefits is a natural and strongly supported function of the private sector."[74]

Representatives from DuPont, Shell Oil, Procter and Gamble, and Monsanto described the key attributes of the method as a means to determine "how to preserve freedom of individual choice."[75] That choice was specifically associated with consumer options, for example, the "choice" of plastic or paper bags.

Cost-benefit analysis provided tools to address political demands for deregulation and free-market economic policies. By contrast, efforts to set low safety standards for chemicals or to interpret the Delaney clause as an absolute-zero standard were cast by industry representatives and many regulatory officials as unscientific, and thus as resulting in overly burdensome regulations and market restrictions. Whether cost-benefit analysis may result in under- or overregulation, whether it fails to fully account for the benefits of protecting public health, whether it overestimates costs to industry, and whether it inadequately prices human life are all questions that have persisted in debates about cost-benefit analysis for decades. Even before Ronald Reagan came into office and slashed regulatory agency budgets, a number of efforts were made to institutionalize the use of cost-benefit analysis in regulatory decision making about chemical risks.

The tension between promoting economic growth and protecting public health came to a head during the administration of Jimmy Carter. Carter came into office during a period of weakening economic conditions, with the strong backing of leading environmental organizations such as the Natural Resources Defense Council and labor unions. Persistent economic stagflation punctuated by a second oil crisis plagued Carter's four years in office. An increasingly well-organized business community that promoted free-market policies, deregulation, and lifting of price controls had gained political strength. The antibusiness days of the late 1960s and early 1970s were over. As David Vogel explains in his classic book on the history of the private sector's political power, *Fluctuating Fortunes*, "Whereas public attention had focused almost exclusively on the inadequacies of business performance during the first two-thirds of the 1970s, by the end of the decade the public had become responsive to many of the complaints of industry about regulatory excesses."[76] By the end of Carter's administration, demands for cost controls in regulation provided strong political support for the adoption of cost-benefit analysis and risk-assessment methods into chemical regulation. In turn, concepts of zero tolerance for carcinogens and health-based standards became associated with big government and the "nanny" state and were subsequently increasingly marginalized as acceptable policy.

While the economic downturn made industry's political complaints more favorably received, the strengthened political power of business was also a result of several years of dedicated organizing. Perhaps the best example of such organization was the formation of the Business Roundtable in 1972.[77] An "informal alliance of about 120 chief executives of the largest and most important corporations in the country," including the chemical and plastics companies and trade associations, the Roundtable functioned "quietly and most effectively behind the scenes; testifying before Congress and in meetings with Administration leaders." Members of the Roundtable operated "as concerned citizens but legally and properly coordinating their efforts." Organized into task groups, the Roundtable focused on tax reform, trade, labor relations, environmental regulations, economic education, wage and price controls, and regulatory reform, with their overarching objective to "assure that business is not working at cross purposes on important issues."[78] Later, in the early 1980s, the Business Roundtable formed its own "Risk, Cost, Benefit Analysis" task force to develop and promote risk-assessment and cost-benefit policies.[79]

While Carter faced a well-organized business community and economic stagnation, he retained a commitment to strong environmental and occupational health and safety, as evidenced by his appointments of progressive leaders to head many regulatory agencies.[80] Eula Bingham, an occupational health and safety expert with extensive research experience in occupational exposures to carcinogens, became assistant secretary of labor for OSHA; Donald Kennedy, a highly respected biologist from Stanford University, was appointed FDA commissioner; and Joseph Califano Jr., former domestic advisor to President Lyndon B. Johnson, became secretary of HEW. Doug Costle, Carter's EPA administrator, came from within the Nixon administration, having formerly served at the OMB and as a member of the Ash Commission, which designed the EPA.[81] These agency leaders struggled to define a consistent policy for chemical carcinogens as part of a broader effort to better coordinate interagency activities. In a letter to President Carter, agency heads outlined a number of interagency initiatives directed toward making the "regulatory processes more efficient for our agencies, for industry, and for the public."[82] Their effort to set a general cancer policy emerged from joint agency efforts to develop similar testing standards, data requirements, and risk-assessment methods; to share information and regulatory standards; and to coordinate research efforts on chemical hazards.

The agencies' varying policies on carcinogens were informed by differences in legal statutes and agency culture and leadership. With strong support from its legal team, Bingham's OSHA interpreted its mandate under the Occupational Safety and Health Act (OSHAct) as setting the lowest possible standards for carcinogens on the basis of the weight of existing evidence. Therefore, if a chemical was deemed a carcinogenic risk, then the standard should be as low as technically feasible as outlined in the law. In this framework, standards were considered health based because they held protection of worker health as paramount in the decision-making process.

The EPA worked to implement risk assessment and cost-benefit analysis as the dominant methods to set standards for all chemicals, including carcinogens, an approach supported by the president's Council of Economic Advisers and the OMB. Using this approach to assess risk allowed safety standards for carcinogens to be set at levels where there was not a significant risk of cancer—for example, where there was a one in a million risk.

But the point at which the risk was minimal might not be same as the lowest level technically possible.

At the FDA, Kennedy inherited a number of highly controversial ongoing regulatory decisions on the safety of saccharin, DES, and the plastic acrylonitrile. Like his predecessors, Kennedy, who after his FDA tenure became provost and president of Stanford and editor of *Science Magazine*, considered the Delaney clause to be redundant and an unnecessary addition to the general safety standard—that is, the FDA did not need the cancer clause to restrict carcinogens.[83] In each of these decisions, the limits and possibilities of the Delaney clause were confronted. While in the 1970s the FDA, like the EPA, had tried to move in the direction of setting risk-based standards for carcinogens, it still had to follow the Delaney clause. Carcinogens ostensibly could be banned because of their hazardous properties alone, but this, while providing the agency considerable regulatory authority, also made evaluating the relative risks of chemical carcinogens difficult. The Delaney clause was a powerful legal means to protect public health reluctantly used by the FDA, as evidenced in the case of cyclamates. But as was proved by the late 1970s with the regulatory decisions on DES and, most notoriously, saccharin, it was at the same time a blunt instrument that for some pointed out regulations' ineffectiveness and inefficiency.[84]

The controversy over the safety of acrylonitrile unfolded over several years, beginning in 1974, when DuPont submitted the first petition for approval of the compound in plastic bottles to the FDA. Although acrylonitrile was approved for use in food-packaging plastics, the FDA responded to DuPont's petition by lowering its safety standard. Several years later, the agency once again lowered the allowable migration level, from 300 ppb to 50 ppb, on the basis of evidence that the chemical caused cancer in experimental animals.[85] Monsanto, which recently had signed a contract with Coca-Cola to begin manufacturing the first plastic soda bottle, resubmitted a petition for the compound's use in such bottles, demonstrating that acrylonitrile migration fell below the new allowable limit and thus was not subject to restriction under the Delaney clause. Yet, in a reversal of past FDA policy, the agency rejected the petition, maintaining that despite the lack of detectable levels acrylonitrile still posed a cancer risk. Monsanto brought suit against the regulatory agency, with legal counsel from Heckman's law firm, Heckman and Keller, and a newly established conservative legal group, the Pacific Legal Foundation.[86] Commissioner Kennedy's lawyers defended the rejection of the petition by using a rather creative argument based on the second law of thermodynamics: when two substances come into contact, diffusion will result. On the basis of this fundamental principle, the FDA lawyers argued that according to the laws of physics acrylonitrile existed, even if it was not within the detectable range, and that it was therefore subject to regulation.

In *Monsanto Co. v. Kennedy* (1979), the D.C. Circuit Court ruled in favor of the industry on the basis that the agency had not sufficiently proved that acrylonitrile was a "food additive" subject to regulation because it had failed to quantitatively demonstrate the migration of the chemical into food. While the judge recognized the creativity of the agency's argument, he concluded that "Congress must have intended the Commissioner to determine with a fair degree of confidence that a substance migrates into food in more than insignificant amounts."[87] The decision upheld the limits of the agency's ability to establish absolute restriction of carcinogens by ruling that the Federal Food, Drug and Cosmetics Act did not intend for the agency to regulate those substances that might migrate into food at insignificant levels defined as below the detectable limit.[88] Again, *de minimus non curat lex.*[89] Such legal consistency in interpretation of the clause explains how PVC stayed on the market in the mid-1970s despite evidence of vinyl chloride carcinogenicity and why the agency ultimately completely restricted DES when very low levels were detected in edible tissue.

When the FDA announced that it would ban saccharin because of evidence of its carcinogenicity, the organized response by industry and the negative public backlash were unprecedented. With cyclamates banned, saccharin was the only synthetic sweetener available on the market, and it was widely used in food and drugs.[90] The considerable emotional response to the proposed saccharin ban, allegedly organized by the Calorie Control Council, provided industry opponents to the Delaney clause ample opportunity to demonstrate to lawmakers the fundamental flaws of the per se standard.[91] The case of saccharin was different in part because it was personal. It touched the heart of America's obsession with weight loss. Industry-led critiques of government overreaching and the need for greater protection of individual and consumer choice resonated with many Americans. Political commentators and news analysts pointed to the saccharin decision as regulation run amok and to the Delaney clause as forcing the agency to make irrational and unscientific decisions—arguments that Edwards and Hutt had made earlier in the 1970s. The seemingly irrational decision-making process required by the Delaney clause was captured in a frequently repeated statement: one would need to drink eight hundred cans of diet soda a week to increase cancer risk.[92]

The agency was stuck in an untenable situation: it had to follow the law while effectively responding to consumer confusion, panic, and frustration. In an effort to uphold the law and respond to concerns voiced by diabetics in particular, Kennedy considered restricting saccharin's use as a food additive but maintaining its availability as an over-the-counter drug. Strong industry objections to the new proposal resulted in a series of congressional hearings and the Saccharin Study and Labeling Act, which issued a moratorium on the saccharin proposal and required continued study of saccharin and a label warning of its risks. Today, saccharin continues to be available as a food additive. Evidence that its carcinogenicity may be limited to bladder cancer in rodents resulted in the removal of hazardous warning labels in 2000.[93]

Like Kennedy, Bingham came directly from academia and took over an agency in crisis. She had studied chemical carcinogens and occupational health at the University of Cincinnati, and her appointment to lead OSHA came with the support of women's groups and unions. In 1973, Bingham spoke out in support of a strike by Shell Oil workers over hazardous working conditions led by Tony Mazzocchi, a prominent leader in worker health and safety, who served as officer in the Oil, Chemical and Atomic Workers Union and was a leading union representative in the passage of the OSHAct in 1970.[94] By winning support from the labor movement as a strong supporter of both workers' right to know and protection from serious hazards in the workplace, Bingham gained the support of the president.

For Carter, improving OSHA's image and work was one of the highest priorities for the Department of Labor, and he personally interviewed Bingham for her position. According to Secretary of Labor F. Ray Marshall, the president had heard complaints about the agency throughout his election campaign. Marshall recalled that Carter felt the OSHAct, which had had so much promise at its inception, had become a "laughingstock."[95] In her interview with the president, Carter gave Bingham two examples of his concerns about the agency's past performance. The first was inappropriate and ineffective citations—in other words, the agency spent too much time on frivolous violations. The second example was the agency's failure to set safety standards to protect workers in the most dangerous settings. The latter example came from Carter's visit to an asbestos plant, where he found the dust level very unsettling and was told by workers that the air quality had been even worse just prior to his visit. Carter believed that OSHA should focus on the "whales and not the minnows." The whales were serious health problems and illness in the workplace, and the minnows were insignificant safety standards.[96] As part of this effort to focus on the "whales," OSHA developed a series of new health regulations for chemical carcinogens, including lead and benzene.[97]

Providing legal support for OSHA's new carcinogen standards was Anson Keller, who previously had served as associate general counsel at the EPA. Keller had resigned from the EPA in protest in 1975 when administrator Russell Train, the agency's second administrator, withdrew his support for the Office of the General Counsel's effort to develop a cancer policy based on no-threshold dose-response principles as articulated by Umberto Saffiotti at the NCI.[98] Keller interpreted OSHA's mandate as setting the lowest feasible safety standards for chemical carcinogens on the basis of the weight of the evidence and the no-threshold principle, which did not include evaluations of costs. Yet Bingham's effort to set standards as low as possible for benzene, lead, acrylonitrile, and arsenic put her in direct conflict with the chemical industry, the OMB, the president's Council of Economic Advisers, and, at times, the EPA.[99]

Beginning in the mid-1970s, the EPA began formulating a cancer policy, supported by administrators Train and Costle, that embraced quantitative assessment of risk. Chemical risk, within this framework, was assessed according to the hazard of the chemical, the exposure to the hazard, and the relationship between the dose and the effect. Train appointed Roy Albert, of the Institute of Environmental Medicine at New York University, to head up the agency's Carcinogen Assessment Group in 1976.[100] Albert strongly supported a quantitative approach to assessing cancer risks. For the regulated industry, this position put the EPA in a favorable light because it provided a process by which exposure to carcinogens could be interpreted as "safe," as opposed to being completely restricted as a hazard per se. After attending a meeting of the plastics trade association in 1978 where he detailed the EPA's policy on carcinogens, SPI president Ralph Harding wrote Albert, saying that Albert's "comments on the approach taken by EPA in dealing with the cancer issue were of great interest to all of us. . . . We look forward to working with you in the future."[101] Risk assessment and cost-benefit analysis were methods supported and indeed promoted by the MCA at a time when removal of the Delaney clause was recognized as politically impossible.[102]

The differences in regulatory approaches to carcinogens came to a head when agency leaders attempted to develop an interagency policy. In 1977, leaders of the major agencies formed the Interagency Regulatory Liaison Group (IRLG) to coordinate research and regulations among the agencies, with a particular focus on establishing a single cancer policy. The

IRLG began as an informal gathering of the agency heads over breakfast and emerged as a cross-agency effort to coordinate research and testing, information transfers, communication, and risk-assessment practices. The group included Doug Costle from the EPA, Kennedy from the FDA, Bingham from OSHA, and John Byington from the Consumer Product Safety Commission.[103]

A final report developed by the IRLG, "Scientific Bases for Identification of Potential Carcinogens and Estimation of Risk," outlined the findings of its working group on risk assessment and attempted to develop a policy to regulate carcinogens.[104] According to Marc Landy, Marc Roberts, and Stephen Thomas's account of the IRLG in *The Environmental Protection Agency*, Bingham objected to original language in the report that she believed too strongly upheld the use of quantitative risk assessment for carcinogens. If the IRLG promoted risk assessment for carcinogens, its recommendations could be interpreted as undermining OSHA's health-based standards for carcinogens. Anson Keller of OSHA worked to bring in Umberto Saffiotti from the NCI to help revise the report. Under pressure from OSHA, the group agreed to include language that emphasized the tremendous amount of uncertainty behind quantitative estimates of cancer risk.[105] In more recent reflections on these events, Bingham noted that she had wanted to use the "emerging approach of RA [risk assessment] but was overruled by agency lawyers."[106] Bingham, like many policy makers, faced not only the conflicting external pressures, viewpoints, and opinions of industry, advocates, members of Congress, and various administrative offices, such as the OMB, but also internal pressures and differences in opinion.

The final IRLG document reflected a compromise. It included a discussion of risk-assessment methods but explicitly highlighted the uncertainties and assumptions of the process. When EPA administrator Costle pushed for the final report's publication in the *Federal Register*, however, according to Landy and his colleagues, Secretary of Labor Marshall refused to sign on. Despite Keller, Saffiotti, and Bingham's support for the final report, Marshall was concerned that printing the IRLG guidelines in the *Federal Register* would create the appearance of accepted regulatory policy, thereby undermining OSHA's no-threshold policy for carcinogens. Costle went ahead and published the report without OSHA's support in 1979; however, subsequent to the report's release, a number of IRLG participants, including Keller, worked to get it peer reviewed.[107]

Industry representatives issued strong criticisms of many issues set forth in the report, including the use of the most sensitive animal models in toxicity testing, the presumption of a linear dose-response relationship, and the report's general tendency to avoid underestimates of risk. Consequently, pressure from private industry, together with the conflicting guidelines issued by OSHA and the failure of the paper to permeate agency policies, rendered the paper moot.

Despite the inability of the agencies to harmonize carcinogenicity policy, and given the fractious nature of concurrent debates at the FDA over the Delaney clause, Bingham was very successful in issuing lowered tolerance limits for a number of carcinogenic substances, including acrylonitrile, asbestos, cotton dust, lead, and benzene. In the case of the cotton dust standard, Carter met personally with Marshall, Bingham, and Charles Schultze, chairman of the Council of Economic Advisers, who advocated a more lenient standard for personal protective equipment. The meeting resulted in Carter directly supporting the Department of Labor's stricter control standard, in opposition to the council's recommendation.[108] In the case of benzene, a major building-block chemical in the petrochemical industry used in products from gasoline to plastics, the agency set the safety standard at the lowest technologically feasible limit, 1 ppm. Because no threshold of safety for benzene had been established, OSHA would set the tolerance limit as low as possible in order to best protect workers' health. A health-based standard, the tolerance limits on benzene exposure in the workplace sought to give greater protection to the worker given uncertainties about benzene risks, exposures, and the costs of implementing new standards.

Not surprisingly, OSHA's success in issuing health-based standards presented a significant threat to regulated industries. In response, the chemical industry established the American Industrial Health Council (AIHC) in late 1977.[109] Working with the AIHC, SPI president Ralph Harding outlined a strategy to counter OSHA's policy that involved lobbying, public relations, and "statements by some of our 'independent' witnesses."[110] For example, the AIHC developed a long list of questions to prepare witnesses prior to their testimony at OSHA's hearing on the benzene standard.[111] The overarching strategy for the industry, according to Harding, was to build pressure on OSHA and the administration by "engender[ing] some skepticism about the doctrinaire positions of the environmentalists"; establishing favorable reporting of industry's alternative plan in the press; instituting letter writing from consumers, small businessmen, and chemical workers to members of Congress about industry's plan; and then sealing the deal with congressional visits from "several constituent groups, to complete the process." This multipronged approach to stopping OSHA's cancer policy, Harding remarked, "has to be many times larger and more effective than anything ever attempted before in the chemical industry."[112]

OSHA's health-based standards for carcinogens represented what Harding perceived as the "emotional and political aspects of the cancer issue."[113] This sentiment held that the public's irrational fear of cancer was resulting in unchecked regulation. When a coalition of public health leaders at the NCI, the National Institute of Environmental Health Sciences, and the National Institute for Occupational Safety and Health issued their "Estimates Report" on the risks of chemical carcinogens in 1978, the report fueled both public anxiety and industry claims that chemical risks were mere hyperbole designed to expand regulation. What historian Proctor calls the "most radically environmentalist U.S. government document ever written," the "Estimates Report" predicted a dramatic rise in environmental cancers due to expanding use of industrial chemicals and called for greater federal efforts in public health prevention.[114] In thirty years, the report predicted, the proportion of cancer mortality attributed to high-production carcinogens such as asbestos and benzene could rise to 40 percent.[115]

At a three-day conference sponsored by the AFL-CIO on Occupational Safety and Health, Secretary of HEW Califano, who endorsed the report's findings, stressed the dire predictions and unabashedly called for greater efforts in health prevention. In a speech before the group of labor leaders and administrative representatives, Vice President Walter Mondale directly confronted economic criticisms of efforts to improve worker safety and health. It's "myopic," Mondale stated, "to argue that programs to protect workers are inflationary."[116] Mondale's remarks were in response to the emerging backlash against environmentalism as a source of economic drag on an already slowed economy. For the chemical industry and the AIHC, the "Estimates Report" merely reflected the extreme positions of environmental advocates. They roundly dismissed its predictions as overblown for political purposes and suggested that the proportion of occupational-related cancers was closer to 1 percent.[117]

The year after its release, the scientific integrity of the "Estimates Report," and consequently the significance of environmental exposures in cancer development, was dealt a serious blow. World-renowned and highly respected epidemiologists Richard Doll and Richard Peto, whose work had contributed to a growing body of evidence supportive of a causal relationship between smoking and lung cancer, published *Causes of Cancer*. The authors attributed the majority of cancers in the United States to tobacco smoke and diet and predicted that occupational exposures accounted for only 4 percent of cancers and pollution only 2 percent. Their findings were widely embraced by government institutes and agencies, as well as the popular press. The toxicity crisis of the decade was drawing to a close. Business interests and concerns about regulation were regaining political legitimacy, while the power of environmental and consumer advocates was waning. Whether intentionally or not, *Causes of Cancer* fulfilled the AIHC's objective to quell public fears of the cancer risks of chemical pollution.[118] Many years later, however, a dark cloud was cast over the integrity of Doll's work when the British newspaper the *Guardian* revealed in 2006 that for twenty years, beginning in 1979, he had been working as a consultant for Monsanto.[119]

Causes of Cancer was part of a significant political sea change in the country. The political Left had collapsed in on itself, and the tide was shifting to the ideological Right. Deregulatory policies accompanied greater institutionalization of cost-benefit analysis and risk assessment within regulatory agencies. Public backlash against the FDA's saccharin decision epitomized a changed landscape of increasing industry organizing and declining public tolerance for expanding regulations. In 1979, just prior to leaving office, Carter signed into law the Paperwork Reduction Act, which established the Office of Information and Regulatory Affairs within the OMB and formalized the OMB's centralized role in reviewing regulatory decisions using cost-benefit analysis. Carter also issued Executive Order 12044, which required the review and revision of all federal regulations to account for cost effectiveness.[120]

Industry's legal challenges to OSHA's health-based standards for carcinogens worked their way through the lower courts, resulting in two major Supreme Court cases in the early 1980s. In *Industrial Union Department, AFL-CIO v. American Petroleum Institute* in 1980, the Supreme Court dealt a major blow to OSHA's authority when it struck down the agency's strict standard for benzene. The court ruled that the agency had failed to prove that benzene presented a "significant risk" to public health at the lowest feasible standard of 1 ppm. The agency's argument before the court reflected its health-based policy approach: "[OSHA] cannot await scientific resolution of the issue but has a mandate to act now . . . and in the absence of . . . no effect levels or safe levels to assume that none exist."[121] The agency argued that the absence of evidence that low levels of benzene are carcinogenic was not the same as the presumption of safety. Given the uncertainty of low-level risks, the agency gave the benefit to workers' health and presumed that no threshold existed for benzene's carcinogenic effect.

The court disagreed with the agency. It ruled that OSHA had failed to meet the requirement, stipulated under the Administrative Procedures Act, that the regulator bear the burden of proving the need for a rule unless statutory language declared otherwise. In other words, the agency had failed to prove the need for such a low standard—it did not demonstrate that benzene presented a significant risk at that standard. In making its decision, the court ruled in favor of the industry's argument that the agency had overstepped its authority in setting a standard without demonstrating a "significant risk" (in which an unsafe workplace condition would be met). However, unlike the lower court's ruling in this case, the Supreme Court stopped short of ruling that OSHA was required to demonstrate economic feasibility through cost-benefit analysis, as the industry petitioners had argued it must do.[122]

The following year, however, the Supreme Court dealt directly with the issue of cost-benefit analysis in the case *American Textile Manufacturers Institute, Inc. v. Donovan.*[123] The case involved several industry petitioners' challenge to OSHA's cotton dust standard due to its failure to conduct a cost-benefit analysis. (In contrast to the benzene case, the petitioners did not question that the agency had demonstrated a significant risk.)[124] The court ruled against the plaintiffs' claim that OSHA must always examine the cost-benefit analysis of its standards, upholding the agency's interpretation of the OSHAct that Congress intended the agency to set technologically achievable health-based standards in favor of workers' health.

These two rulings from the high court provided a framework in which OSHA could set standards in the future: the agency must conduct quantitative risk assessments to determine significant risks and then determine the technological feasibility of a given standard.[125] Given these rulings and growing frustration with the limits of the Delaney clause, it was clear by the end of the 1970s that carcinogens would not be regulated as hazards per se and that consequently markets for PVC, benzene, and other known carcinogens could continue to grow.

By the late 1970s, public fears of chemical pollution and environmentalists' demands for greater regulatory authority over chemicals were dampened by arguments that excessive regulation was strangling the free market, driving inflation, and dampening economic growth. The Delaney clause was "out" and quantitative methods of risk assessment were "in." Risk assessment emerged as the practical response to politically charged disputes about how to manage or completely restrict carcinogens. In the coming years, however, risk assessments of highly controversial compounds—

most notoriously the decades-long process to evaluate dioxin, resulting in a document now well over a thousand pages long—would demonstrate the unavoidable political morass that assessing chemical risk and safety could become.

Risk-assessment practice in the 1980s continued to maintain two models of dose-response relationship: one for carcinogens and another for all non-carcinogens. This distinction reflected heightened public concern about cancer risks and the effort to build conservative protections for public health into the models. The dose-response relationship for noncarcinogens was assumed to stop at a certain threshold: even if detection limits continued to drop, the status of a chemical's safety could remain stable provided the presumed exposure level fell below the toxic threshold. And indeed, detection limits dropped significantly, from 20 to 100 ppm in the late 1950s to parts per *billion* by the early 1970s. Today detectible levels are reported in parts per quadrillion.[126]

In the case of chemical carcinogens, risk-assessment models assumed a linear dose-response model. This sparked debates about the meaning of zero and whether a safety standard could be set at some level above absolute zero—as either a risk-based standard or according to the detection limit. As in the case of the Delaney clause, zero was defined by the detection limit. This meant that as technologies advanced, the presumed safe use of a carcinogen could be overturned, as was the case with DES. The Delaney clause, therefore, opened the legal door to ban potentially dozens of chemicals on the market, as exemplified by the cyclamate and DES bans. From industry's perspective, this was the crux of the problem with the assumption of a linear dose-response for carcinogens, and with the Delaney clause itself. How could a safety standard be set if one assumed that carcinogens presented significant risks at any level above absolute zero?

Amid all the debate in the 1960s and 1970s over how to regulate chemical carcinogens and interpret the dose-response relationship, an important concern was lost. The original arguments for removing carcinogens from the food supply, articulated by Hueper and the International Union against Cancer in the 1950s, focused on exposure in the vulnerable developing young and the delayed risks of even low-level exposure. That estrogenic compounds could be both carcinogenic and low in toxicity simply could not be explained by the difference in dose-response curves for carcinogens and noncarcinogens. An important element—the timing of exposure—was missing from the model of risk. The standard study protocol for testing carcinogenicity involved chronic exposure to high doses of a suspected chemical in adult rodents. Using these data to determine population risks required extrapolating effects at high doses to low doses according to the dose-response curve, as well as presuming that the risks of exposure in adulthood were equivalent to the risks of exposure in early development.

Rhetorical arguments against the Delaney clause and health-protective standards highlighted the inconsistency of this approach to evaluating low levels' carcinogenic risks. The best example was the suggestion that one would need to drink eight hundred cans of diet soda to develop cancer from saccharin, and this proved to be an effective image to demonstrate the significant limitations of the Delaney clause. Similarly, the "single rat" argument meant that one must make an unscientific decision given any evidence of risk. Absent from these simplified extrapolations of risks were considerations of the different exposures and effects as they occurred in pregnant woman and young children. The idea articulated in the early 1950s—that babies are not just little adults—was lost.

Throughout the late twentieth century, as researchers shifted their gaze to study health effects from the lower and lower concentrations of chemicals increasingly being detected in the environment and the human body, new understandings about the relationship between low-level exposure and disease risks emerged. Particular attention began to be paid to the timing of exposure and the vulnerability of fetuses, infants, and children. In part, continued research on DES forged new insight into how some chemicals interact with hormones and biological development to create adverse health effects. As research priorities changed, so too did the standard assumptions about dose-response relationships at the core of chemical policy. Some argued for harmonizing the linear and nonlinear models; others argued for additional dose-response models. A profound gap began to emerge between scientific research on industrial chemicals that pointed to biological effects at lower and lower concentrations, and the regulatory process to define chemical safety that broadly presumed safety at low levels of exposure. Public recognition and awareness of this gap, and frustrations about the ineffective regulatory system, provided fodder for yet another political crisis, this one in the 1990s, which destabilized the meaning of chemical safety in an age when we had all become "a little plastic."

3   Regulatory Toxicity Testing and Environmental Estrogens

David Rall, a cancer specialist and physician, institutionalized the field of environmental health sciences. For nearly two decades, from 1971 to 1990, he served as the second director of the newly established National Institute of Environmental Health Sciences (NIEHS). Under his leadership, the institution was transformed from a small group of researchers to an expansive, sprawling tri-city research hub known as Research Triangle Park, set amid the pine trees of central North Carolina. He led efforts to ground the study of public exposure to chemicals and other environmental hazards in the basic sciences, hiring researchers trained in cellular and molecular biology, genetics, reproductive biology, and physiology.[1] Rall sought to implement the original mission of the federal institute as outlined in 1964: "To determine the magnitude and significance of the hazard to man's health inherent in long-term exposures to low-level concentrations of biological, chemical, and physical environmental agents . . . to identify the underlying mechanisms of adverse response . . . to set standards and to provide predictive guides to be used by control agencies for protective or preventive measures."[2] That same year, Wilhelm Hueper retired from the National Cancer Institute (NCI), ending his controversial tenure. The NCI subsequently closed its Environmental Cancer Section and dispersed Hueper's extensive library, effectively erasing his work from its institutional memory.[3]

Read carefully, the NIEHS mission outlines two federal responsibilities for managing and mitigating environmental hazards. The first is the production of exploratory research directed at elucidating the "magnitude and significance" of human exposure to environmental hazards, including the biological mechanisms by which such hazards elicit adverse effects. The second explicitly directs the development of standards for regulatory or "control agencies." Over the course of Rall's long tenure at the NIEHS, these responsibilities guided the development of two distinct, yet overlapping (and at times conflicting), processes to produce information about a chemical's risk and the means to define its safety.

Simply put, these two approaches to studying chemical risks can be identified as hypothesis-driven, exploratory environmental health research, and regulatory toxicity testing for safety standards. Environmental health research and regulatory toxicity testing both produce data and information on chemical hazards and risks, but this does not always mean that the knowledge gleaned from each approach is consistent with the other's. These approaches often differ in the questions they pose, the tools they use, the assumptions (their means to understand risk) they make, and the ends to which they put information. In general, regulatory toxicity tests are designed to determine whether well-defined adverse toxic effects are observed, and at which doses, in order to set a safety standard for regulatory purposes. Demand for toxicity testing increased as a result of a number of environmental laws passed in the 1970s that gave the newly established EPA the authority to set standards for chemical hazards, including the passage of the Safe Drinking Water Act in 1974 and the Toxic Substances Control Act (TSCA) in 1976. Conducted by private testing firms that employed standardized methods, regulatory toxicity testing became increasingly rationalized and professionalized.

As regulatory testing expanded, so too did research in hypothesis-driven environmental health conducted at the NIEHS and in academic laboratories with federal funding from the NIH. In contrast to regulatory toxicity testing, environmental health research is a hypothesis-driven discipline in which the intent is not to use standard testing practices to determine a set standard of safety based on a specific set of endpoints but to test theoretical assumptions, to explore paradoxical findings, and to expand understanding of how and why a chemical might interact with various biological processes. The questions asked, tools used, and assumptions made can vary widely, given that academic researchers often seek to generate novel findings for publication and professional survival. A researcher can practice excellent, decent, or shoddy work regardless of the model he or she is working under. In other words, neither model represents a better science or a good science per se, but different information can emerge from each. When this occurs, scientific and political conflict can ignite around the questions of whose science and which research determines a chemical's safety.

An important disjuncture between regulatory toxicity testing and environmental health research began to emerge in the 1970s and 1980s as scientists tested environmental estrogens. For decades, beginning with the mass marketing and production of estrogens, including synthetic estrogens such as DES and hormone replacement therapy, researchers and regulators have debated whether estrogens are carcinogenic and have produced conflicting evidence of both preventative and adverse effects. After the first epidemiological report of increased vaginal cancer risk from fetal exposure to DES was published in 1971, research on the risks of estrogens increasingly began to consider other compounds with estrogenlike properties, including industrial chemicals.[4] Researchers at the NIEHS, notably John McLachlan, investigated developmental effects of fetal exposure to environmental estrogens, such as DDT and PCBs. This work drew on the emerging evidence of carcinogenic effects of fetal exposure to DES and research on the developmental effects of transplacental (from mother to fetus) exposures to environmental factors such as tobacco smoke and pharmaceuticals.

The lessons that began to emerge from research on transplacental exposure to environmental estrogens, which would become more clearly articulated as part of the thesis of endocrine disruption by the 1990s, were twofold: first, the timing of exposure strongly influences the effect of the chemical; and, second, if exposure occurs during early development, very low levels of exposure can negatively affect reproductive growth and development later in life. Within the regulatory toxicity model, however, neither factor became an integral part of the testing process. For instance, standard carcinogenicity tests used adult or juvenile animals and exposed them to high doses of chemicals. Hueper's argument that some compounds such as estrogens could be carcinogenic but of low toxicity was absent from these tests' organizing logic. Similarly, developmental and reproductive toxicity tests, which allowed some consideration of the timing of exposure, assumed that the dose was the principal factor in determining the effect, and these tests used very high doses to test for risks of serious mutations, birth defects, and infertility. The distinction between lessons emerging from estrogen research and regulatory toxicity testing methods and assumptions is evident in the research on BPA conducted in the late 1970s and early 1980s.

Produced in high volumes for plastics production, BPA came under scrutiny by the NIEHS because of its commercial success and related potential for widespread exposure. Beginning in the late 1970s, the NIEHS coordinated a number of studies, including developmental and reproductive toxicity tests and a high-dose carcinogenicity study, that laid the foundation for the establishment of a safety standard for BPA in 1988. The chemical's estrogenic properties, though recognized, were framed as secondary to concerns about its toxicity and carcinogenicity. As a result, regulatory tests provided guidance on BPA's safety, while simultaneously research on estrogenic chemicals raised new questions about the risks of long-term, low-level fetal exposure to it. By the late 1970s, the use of DES was banned because of carcinogenic risks, and estrogen's therapeutic use in menopausal women also came under serious scrutiny when two studies found increased endometrial cancers associated with use of the drug.[5] For the next several decades, scientists, regulators, and environmental advocates would argue over how to evaluate the risks of environmental estrogens and whether the regulatory toxicity testing process adequately accounted for these biological mechanisms. The roots of the contemporary debate over BPA safety, therefore, lay in the 1970s and 1980s with the expansion of both regulatory toxicity testing and environment health research.

AN INTRODUCTION TO BPA IN PLASTICS AND AS AN ESTROGEN

After it entered commercial production in the 1950s, demand for BPA grew rapidly. It first found valuable application in the manufacture of epoxy resins widely used as adhesives in flooring, water main filters, and construction, and as protective coatings in metal drums, reinforced pipes, and cans, including food cans. Later applications included the production of polycarbonate, tetrabromobisphenol A (used as a flame retardant), unsaturated polyester, polysulfone, polyetherimide, polyarylate resin, and polyester-styrene plastics.[6] As with thousands of other industrial chemicals in production after the end of the Second World War, there was limited knowledge of BPA's toxicological effects. How BPA's rapid commercial success affected worker and public health was virtually unknown, even as production expanded dramatically.

No regulatory standard for a safe exposure level to BPA existed until 1988. BPA was, however, regulated by the FDA after the passage of the 1958 Federal Food, Drug and Cosmetics Act. Since BPA was "reasonably [to] be expected" to migrate from epoxy resins, used since the 1950s to replace tin in the lining of metal food cans, the chemical fell under the regulatory oversight of the FDA as an indirect food additive.[7] The FDA has permitted the use of BPA in adhesives; resinous and polymeric coatings (including the epoxy resins used to line food cans); polyolefin coatings; BPA-epichlo-rohydrin (EPH) resins and thermosetting epoxy resins; and rubber articles intended for repeated use; it has also permitted several uses of BPA-EPH resins and adjuvants (chemicals used to modify the effects of other agents), often in drugs and vaccines.[8] Once the FDA approved a use for BPA as an indirect additive, any manufacturer could use the product without notifying the agency. Without regulatory accountability, the agency lost the ability to determine all the various uses of BPA in the marketplace—a process made even more difficult with the rapid proliferation of plastics and BPA production. As the FDA posted on its website in 2010: "Today there exist hundreds of different formulations for BPA-containing epoxy linings, which have varying characteristics. As currently regulated, manufacturers are not required to disclose to FDA the existence or nature of these formulations. Furthermore, if FDA were to decide to revoke one or more approved uses, FDA would need to undertake what could be a lengthy process of rulemaking to accomplish this goal."[9] As with most indirect food additives, including other forms of plastic, such as PVC and acrylonitrile—both permitted after high-profile debates in the mid- to late 1970s—the FDA has long presumed that the low doses of migration present insignificant risks to public health. For decades of its use in food packaging, however, knowledge about BPA was very limited. Yet as production expanded, so too did research on potential health risks to workers and the general public.

A very basic understanding of BPA's toxicity was outlined in a voluntary safety review issued in 1967 by the American Industrial Hygiene Association (AIHA), an organization established in 1939 by leading industrial manufacturers. BPA was said to result in skin and eye irritation, hypersensitization, and allergic responses from inhalation, skin contact, and ingestion, and to be rapidly metabolized in the adult rodent.[10] Protection from such workplace hazards as directed by the association involved standard industrial safety practices: basic hygiene and protective equipment, such as gloves. Long-term chronic exposure was presumably of marginal concern to producers because of the evidence of rapid metabolism in rodents.[11]

Overlooked in the 1967 review was BPA's "marked oestrogenic action," as it was described in a plastics textbook published the following year.[12] BPA's estrogenlike activity had been known since Dodds conducted his estrogen research in the 1930s. While Dodds passed over BPA as a possible pharmaceutical because of its weaker estrogenicity as compared to DES, chemists in laboratories in Switzerland and the United States found a potential commercial use for the chemical in the late 1930s. Both laboratories created a BPA-based epoxy resin, a strong adhesive material.[13] Beginning in the 1950s, epoxy resins were commercially produced for two main purposes: first, as an additive in paints for metal coatings; and second, in dental restorative work as an adhesive, for example in root canal work, adhering caps, and tooth bonding and filling.[14] Epoxy resins were used in industrial applications to prevent metal corrosion and to extend the life of a wide array of metal tools and machines, from household appliances, cars, dairy equipment, and office equipment to tools. As a protective sealant, epoxy resins were excellent alternatives to tin in lining metal cans, including food, beverage, and motor oil cans.[15] Lining all sorts of steel piping with epoxy resins also became uniform practice by the early 1960s.[16] Commercial production of BPA for use in epoxy resins reached twenty-five million pounds per year in the United States by the mid-1950s.[17]

Then, in 1957, Bayer and General Electric simultaneously announced a new polycarbonate plastic that exhibited promising properties of strength, clarity, and heat resistance.[18] Polycarbonates are repeating chains of BPA joined together by chemical bonds—in other words, polymerized BPA. Mobay, a joint venture between Monsanto and Bayer, and General Electric, under a cross-licensing agreement with Bayer, began commercial production of polycarbonates.[19] In 1959, General Electric announced plans to build a new "multi-million dollar" facility near Mount Vernon, Indiana, along the Ohio River, to produce its new Lexan polycarbonate for "military aircraft and guided missile parts."[20] The strength of polycarbonates put them in a class of plastics called engineering resins, plastics characterized as being tough enough to replace steel. Lightweight like most plastics, polycarbonates could now replace metal and glass in cars, planes, military equipment, and industrial parts, opening up vast new markets.

By the late 1970s, when BPA came under the scrutiny of government researchers, polycarbonate production was still expanding to meet new markets, despite the economic recession and oil shocks of the decade. Polycarbonate plastics found new markets in high-impact car bumpers, safety and optical glass, and baby and water bottles.[21] Between 1977 and 1986, demand for polycarbonate resin in the United States increased 114 percent and production rose from 161 million pounds to 345 million pounds.[22] This explosion in plastics production marked the beginning of the "plastics age"—when plastics production overtook steel.[23] With such tremendous market growth in plastics, BPA production reached unprecedented heights in the 1980s. By 1985, close to a billion pounds of BPA were produced in the United States alone, with continued growth expected.[24] But was the production of hundreds of millions of pounds of an estrogenic compound safe?

REGULATORY TESTING FOR CARCINOGENICITY
IN THE LATE 1970S

When the TSCA went into effect in 1977, the EPA began to accumulate data on all industrial chemicals in commerce for the first time. Its first official compilation of chemicals in production listed sixty-two thousand compounds. Of all the thousands of chemicals in production and in need of testing, the NCI selected BPA for an expensive two-year carcinogenicity study as part of its Carcinogenesis Bioassay program "because of widespread occupational and consumer exposure to the substance and because no other studies had been done."[25] In the United States at the time, over half a billion pounds of BPA were produced by a number of leading petrochemical companies, including Dow Chemical Company, Shell Chemical Company, Union Carbide, and General Electric. Despite the very high volume of production, little was known about BPA, as noted earlier, save for a few toxicity studies conducted by the producing companies.

Government-sponsored testing of BPA safety began in earnest with the initiation of the NCI's carcinogenesis study in the late 1970s, followed by several reproductive and developmental toxicity tests in the early 1980s. Though initiated and paid for by the government, the research itself was conducted by private contracting laboratories. The first of these studies was a long-term carcinogenesis study initiated by the NCI's Carcinogenesis Bioassay Program and conducted by a private testing firm, Tractor-Jitco.[26] The company managed all of the carcinogenesis tests for the NCI at the time and subcontracted much of the work to a number of private testing firms.

Since its establishment in 1971 as part of the National Cancer Act and Nixon's war on cancer, the NCI's carcinogenesis program had suffered from chronic staff shortages and assessment backlogs as government funding priorities focused on cancer treatments rather than prevention. By the early 1970s, the program relied heavily on its main contractor, Tractor-Jitco, to coordinate assessments through subcontracting agreements.[27] Many private toxicological facilities sprang up around the country in response to the EPA's new testing needs. For instance, in anticipation of the passage of the TSCA in 1976, the chemical trade association established the Chemical Industry Institute of Toxicology, not far from the campus of the NIEHS.

"[Toxicity] testing has become big business for industry and for the government," wrote Harold Stewart, then consultant to the NCI, to his longtime colleague and friend Hueper in 1973. "Around here things are not like they used to be when you and I each set up own experiments and did our own autopsies. Now at NCI it's all contract work." Stewart continued presciently, "One of the weak spots is the monitoring of the government contracts."[28]

Critical lapses in federal oversight of private contracting firms came to national attention during a federal investigation of the largest private testing facility in the United States, Industrial Bio-Test (IBT), beginning in the mid-1970s. A subsidiary of Nalco Chemical Company, IBT conducted hundreds of safety tests for pesticide companies and other chemical companies. Led by the EPA and FDA, the investigation of IBT found appalling, unsanitary laboratory conditions, abysmal recordkeeping, and noxious gases—conditions that unequivocally undermined the credibility of the laboratory and any testing results generated there. As Gerald Markowitz and David Rosner reveal in *Deceit and Denial*, "MCA's [the Manufacturing Chemists' Association's] own investigators learned in 1979 that IBT's research on vinyl chloride's effect on rats, mice and hamsters was so flawed that 'the study by IBT is scientifically unacceptable.'"[29] Whether the conditions and recordkeeping practices at the lab were a result of negligence or deliberate tampering became a focus in a criminal investigation that ended in 1983, when three men from IBT were found guilty of fraud (specifically doctoring of data) and sentenced to prison. The fraudulent practices of IBT brought into question 15 percent of the pesticides approved for use in the United States by that point. The EPA demanded that 235 chemical companies reexamine over four thousand tests previously conducted by the laboratory.[30]

The terrible conditions and poor practices at IBT were an embarrassment to the credibility of both the chemical industry and its regulators. Dozens of chemical companies—knowingly or not—had relied on a fraudulent company to test the safety of their products, and the agencies charged with oversight responsibility had been found asleep at the wheel. The system for assessing chemical safety was a mess. In response to the scandal, and to demonstrate to the public and lawmakers that the agencies had some modicum of control, the FDA and EPA, together with the chemical industry, established standards for "good laboratory practices" (GLP),the FDA in 1978 and the EPA in 1983.[31] These outlined federal rules for conducting research on the health effects or safety of drugs and chemicals for regulatory purposes. The GLP standards established specific guidelines for proper care and feeding of laboratory animals, facility maintenance, accurate and reliable calibration and care of equipment, collection and storage of raw data, and inspection requirements.[32] Assuming that a study complied with GLP guidelines ostensibly would provide the federal government and private industry with some assurance of the quality of an experiment.

Investigation of the IBT facility brought the lack of federal oversight of private contractors to national attention. In 1979, Congressman Henry Waxman (D-CA) seized on the issue and ordered the U.S. General Accounting Office (GAO, now called the Government Accountability Office) to investigate the private contracting facilities used in the NCI's Carcinogenesis Bioassay program. Waxman wanted to know whether the problems of private contractors and the failures of government oversight extended to this under-resourced testing program as well.

The GAO's investigation revealed a number of serious problems with both the federal institute's oversight of the main contractor, Tractor-Jitco, and conditions in some of subcontracting labs. One subcontracting facility in particular, Litton Bionetics, received the lowest rating in the GAO report. Investigators found cracks in its walls, floors, and ceiling; poor quality-control measures (such as the testing of animals with different chemicals in the same room); maintenance of quarantine animals (those not yet in a study) in the same room with test animals; and failure to provide pathologists with all available data on the animals. Any one of these poor conditions, the GAO report concluded, could have affected research findings.[33] GAO also revealed that Tractor-Jitco was well aware of the problems at Litton but continued to issue it government contracts. After visiting the Litton facility in 1977, the director of the NCI's Carcinogenesis Bioassay Program recommended to Tractor-Jitco that no new long-term cancer studies be initiated with Litton.[34] And yet, despite these recommendations, Tractor-Jitco maintained its contract with the government as the program was transferred to a new federal institution, the National Toxicology Program (NTP), and Litton received the contract for the carcinogenesis study of BPA.[35]

The carcinogenesis study of BPA at Litton was ongoing during the laboratory investigations, beginning in 1977 and finishing in 1979. The protocol involved exposing two species of male and female rodents, rats (strain F344) and mice (strain B6C3F1), daily to one of two BPA dosing levels in their food or a control for two years. This design followed the standard protocol for the carcinogenesis testing program, which has changed little in the past thirty-five years.[36] The doses used were selected from the results of two short-term studies: an acute, single high-dose study and a subchronic fourteen-day and ninety-one-day feeding study. Male and female rats were fed 0, 1,000, or 2,000 ppm; male mice were fed 0, 1,000, or 5,000 ppm;

and female mice were fed 0, 5,000, or 10,000 ppm.[37] The results were to be handed over to the NCI, which would then issue a final report on the carcinogenicity of BPA. But while the rats and mice ate their BPA-laced food, responsibility for the testing program was transferred to the NTP.

Joseph Califano, secretary of Health, Education and Welfare (HEW) under President Carter, established the NTP to assist in greater coordination and expansion of toxicity testing. The new institution was a response to the ominous predictions of rising cancer rates outlined in the NCI, the NIEHS, and the National Institute of Occupational Safety and Health (NIOSH) "Estimates Report" in 1978 (see chapter 2) and mounting economic and political pressure to increase government efficiency. The NTP was designed to reduce fiscal inefficiency by coordinating toxicity testing across the various federal agencies and institutions. Its budget draws from HEW (now Health and Human Services), and includes participation from the FDA, the NCI, the NIOSH, and the NIEHS, but its home is at the NIEHS. As part of this new initiative to centralize the coordination of research and improve federal oversight of toxicity research, the NCI's program for testing carcinogens was transferred to the NTP.

NIEHS director Rall, who also oversaw the NTP, was well aware of the oversight and contracting problems that plagued the Carcinogenesis Bioassay program at the NCI. During a Senate hearing on federal oversight of the carcinogenesis testing program not long after it had been transferred from the NCI to the NTP, Rall told Senator Al Gore: "We have been very, very concerned about the problems with Tractor-Jitco and we shared NCI's concerns and completely agreed with the NCI decision in May of 1979 to phase out the prime contractor."[38] Rall assured members of Congress that the NTP was taking steps to improve the quality and oversight of testing and to expand the tests conducted on chemicals.[39] One of these efforts included subjecting carcinogenesis studies to independent peer review. Indeed, when the results of the BPA study came in to the NTP, an independent review committee was established as part of the preparation of a final report.

Regardless of the facility problems reported at Litton at the time of the BPA study, the results were kept, and in 1980 the NTP's Board of Scientific Counselors and Panel of Experts conducted a peer review of the findings. By this time, the principal investigator for the BPA study at Litton Bionetics, Dr. E. Gordon, had taken a position with Mobil Chemical Corporation.[40] The final report, issued by the NTP in 1982, concluded that there was "no convincing evidence of carcinogenicity." The clarity of this statement, however, did not fully reflect the ambiguities in the study's findings. Several members of the review panel insisted that the NTP report qualify the statement that "bisphenol A is not carcinogenic" by including "the facts that leukemia in male rats showed a significant positive trend, that leukemia incidence in high-dose male rats was considered not significant only on the basis of the Bonferroni criteria [a statistical test], that leukemia incidence was also elevated in female rats and male mice, and that the significance of interstitial-cell tumors of the testes in rats was dismissed on the basis of historical control data," as opposed to comparison with the experiment's control animals.[41] Additionally, when the rat data were reviewed separately from the mice data, the report concluded: "The evidence is suggestive of a carcinogenic effect on the hematopoietic [blood cell] system."[42]

When the committee reviewed the BPA study results, the NTP had only two categories for evaluating cancer data. There was either "convincing evidence" or "no convincing evidence." This sharp distinction left little room for ambiguous or suggestive, yet inconclusive, findings that the committee attempted to capture in a qualifying statement. Not long after the BPA report, the NTP adopted a multi-tiered classification scheme to sort evidence of carcinogenicity. The new scheme was similar to one first established by the International Agency for Research on Cancer in 1965 that included five levels of evidence.[43] In 1982, however, the bright line between "convincing" and "no convincing" evidence allowed the EPA to interpret the NTP's conclusion as proof that BPA was "noncarcinogenic." All the ambiguity and qualifications associated with the findings disappeared, along with any early-warning indications about exposure to the chemical. BPA was on its way to becoming officially "safe."

DETERMINING THE SAFETY OF BPA

Faced with the monumental task of evaluating the tens of thousands of chemicals already on the market when the TSCA passed in 1976, the EPA focused on those chemicals in high-volume production, such as BPA. After the NTP released its report on BPA in 1982, the EPA conducted what was called a preliminary information review of all existing research on the chemical as part of its authority under the TSCA. With the acknowledged support of the Dow Chemical Company in providing data, the EPA's preliminary information review on BPA, led by George Parris, detailed toxicological research, environmental fate and transport (i.e., the movement of BPA in the water, air, and soil once the compound was released into the environment), occupational exposure, and basic production information.[44] The report made a number of critical conclusions about BPA's carcinogenicity and toxicity. First, the report declared the chemical to be "noncarcinogenic." Thus the NTP statement of "no convincing evidence" was inappropriately conflated with the more matter-of-fact category of "noncarcinogenic." Second, BPA was determined to be nonmutagenic. Although there was evidence of mutagenicity in epoxy resins, industry studies of BPA confirmed to the EPA that this chemical was not the guilty culprit.[45] And third, the report detailed evidence of BPA's estrogenicity from a 1970 paper and a small study conducted at NIOSH in 1981.[46] According to the preliminary information review, BPA's estrogenicity was more potent than DDT's and might be the "mechanism by which implantation [of fertilized eggs] was blocked in rats" in the NIOSH study. Yet because BPA "does not have the half-life that DDT has, and it does not bioconcentrate like DDT," the reported concluded, "it may not pose a threat to viability under probable environmental conditions."[47] Without compelling evidence of carcinogenicity, mutagenicity, or teratogenicity (that is, its ability to cause birth defects), the toxicological research on BPA in 1982 fell far short of presenting the "unreasonable risk" necessary for the EPA to issue any regulation under the TSCA.

The evaluation of BPA's estrogenicity in the 1982 EPA review was based on the few small studies publicly available at the time. These included a 1970 paper by Joel Bitman and Helene Cecil, researchers at the Animal Husbandry Research Division in Beltsville, Maryland. Bitman and Cecil identified a number of synthetically produced estrogenic compounds, including different analogs (structural arrangements of the same chemical compound) of DDT. In an earlier paper, they reported that ortho, para DDT, or *o,p'*-DDT, was the most estrogenic analog.[48] Bitman and Cecil's research built on the work of Dodds and a number of other researchers who had identified various chemical structures with estrogenic potential. (They extensively cited a 117-page review of the chemical structures of estrogenic compounds published in 1945 and a 1950 study of the estrogenicity of DDT in cockerels.)[49] In their exploration of chemicals with two-dimensional structures similar to natural estradiol, Bitman and Cecil, like Dodds, identified the notable estrogenic activity of BPA along with several different structures (congeners) of PCBs. BPA, they reported, was "as active as *o,p'*-DDT."[50]

This basic research on BPA's estrogenicity reflected the small but growing interest in the hormonal activity of industrial chemicals as research agendas at the NIEHS and NIOSH expanded beyond chemical carcinogenicity in the late 1970s and early 1980s. In comments submitted

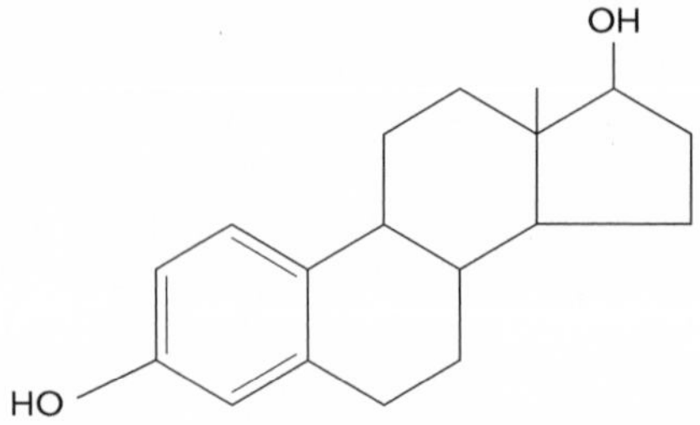

Estradiol

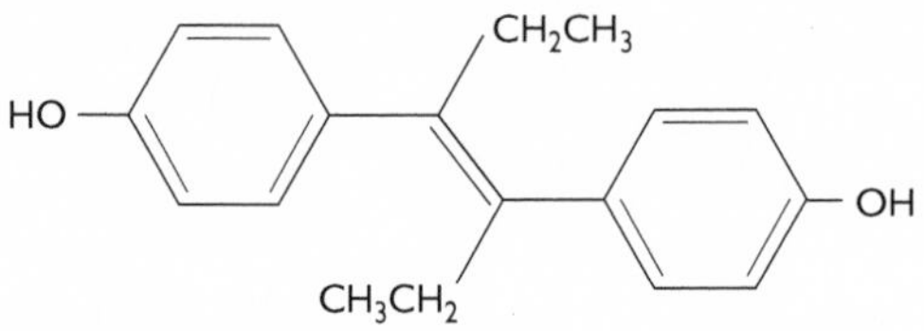

Diethylstilbestrol (DES)

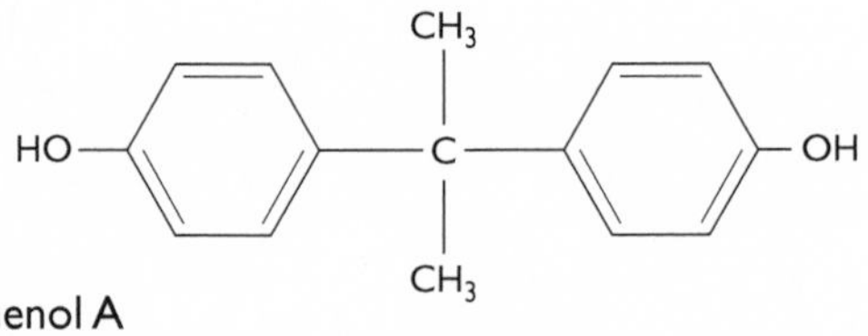

Bisphenol A

Figure 3. Two-dimensional drawings of estrogenic compounds.

to members of Congress in 1981, NIEHS director Rall outlined the priorities of the institute: "We want the final result not to be a mere statement that a chemical does or does not cause cancer in animals, but a complete toxicological profile."[51] This expansion of the investigation of a chemical's toxicity influenced research priorities at NIOSH as well. Researchers in the field of occupational health began to consider reproductive effects of chemical exposures as a result of findings from a number of high-profile accidents. For instance, in the mid-1970s, a NIOSH investigation of an accident at a pesticide factory in Hopewell, Virginia, where workers were exposed to high levels of kepone, found high rates of sterility and loss of libido, in addition to tremors, irritability, and memory loss, in exposed male workers. This NIOSH report triggered further research on the pesticide in 1980 that demonstrated its ability to mimic estrogen.[52] Additional investigations of reproductive hazards in the workplace during the mid- to late 1970s included studies on lead, PCBs, vinyl chloride monomer, and a number of pesticides known to cause infertility and low sperm count in men. Research in 1974 found increased rates of stillbirths and miscarriages in the wives of workers exposed to vinyl chloride monomer.[53] In 1979, researchers reported a correlation between exposure to PCBs, banned in the United States in 1976, and declining sperm counts.[54]

"Recently," NIOSH researchers wrote at the beginning of a small pilot study of BPA, "it has become increasingly clear that occupational exposure to industrial chemicals may impair functional reproductive capacity or may affect the fetus."[55] BPA's estrogenicity and its high-volume production meant that large numbers of workers were potentially exposed. In 1980, NIOSH estimated that 38,079 workers were potentially exposed to BPA; for the years 1981–83, the National Occupation Exposure Survey estimated that 92,138 workers were potentially exposed to BPA, of whom 15,144 were women.[56] The significant number of potentially exposed workers and BPA's estrogenicity contributed to its selection for a small pilot study exploring the reproductive and developmental effects of a number of occupational hazards.

The NIOSH study involved a standard strain of rat called Sprague-Dawley, used in reproductive toxicity studies at the time and bred for its large litter size and robustness (and later recognized as being relatively insensitive to estrogen).[57] Initially, the researchers exposed pregnant female rats to 125 milligrams of BPA per kilogram of body weight (mg/kg) early in pregnancy, from days 1 to 15. When it became clear that very few of the exposed rats were able to sustain their pregnancies, four additional pregnant rats were added to the study and the dose was lowered to 85 mg/kg. The researchers reported that exposure to BPA at the high dose (125 mg/kg) impaired pregnancy and that both doses reduced the number of live pups born. These were the effects noted in the EPA's review. In those pups that did survive to birth, the researchers reported a significant increase in birth defects, including "incomplete skeletal ossification" in the 85 mg/kg treatment group and a significance increase in animals with "enlarged cerebral ventricles or hydrocephaly [buildup of fluid in the brain]" in the 125 mg/kg group. The infertility observed in the exposed animals was attributed to the estrogenicity of BPA. In light of the severity of the birth defects, the authors recommended that a larger, more statistically powerful study be conducted.[58]

Limited budgets and staffing, however, cut ongoing reproductive and developmental toxicity testing at the NIOSH. When Ronald Reagan took office, his administration froze the institute's budget.[59] The political and fiscal marginalization of the NIOSH along with OSHA in the early 1980s strongly curtailed investigations of occupational exposures and the growing interest in reproductive health effects of chemical exposure in workers. Lack of funding and staff also made some of the collaborative efforts among research institutes promoted during President Carter's administration increasingly difficult. Yet BPA did prove to be something of an exception. Findings from the NIOSH's pilot study on BPA did successfully initiate several larger studies of its reproductive toxicity and mutagenicity by the NIEHS in the mid-1980s as a result of collaborative efforts among researchers at various institutes within the NTP.[60]

At the NIEHS, James Lamb, who directed reproductive toxicity testing, initiated several reproductive and developmental toxicity studies of BPA in the mid-1980s. The objective of the reproductive toxicity study, conducted under contract with the private firm Research Triangle Institute, was twofold. First, it would "evaluate and refine this test system" (the newly developed "Fertility Assessment by Continuous Breeding"), and second, it would generate "data on chemical toxicity."[61] BPA was a guinea pig of sorts for the new testing protocol—a process that would also provide some important data on reproductive toxicity. The final report issued by the NTP on BPA's reproductive toxicity in 1985 concluded that BPA was indeed a "reproductive toxicant" on the basis of evidence that it impaired fertility and "reduced sperm motility and weight of some male reproductive organs." Given that these effects were observed in conjunction with serious liver and kidney toxicity, the researchers concluded that the reproductive damage might have been a secondary effect of more general toxicity.[62] In other words, the doses used were so high that the reproductive effects might have resulted from generalized toxicity as opposed to the chemical's estrogenic effects.

A second NTP study sought to explore the potential teratogenic effects of BPA: that is, its ability to cause birth defects, as observed in the NIOSH study. The NTP teratogenic test altered the design of the NIOSH study in two important ways. First, the dosing levels used were much higher than those in the NIOSH test, in order to increase the likelihood of observing adverse effects. Second, the timing of the exposure was shifted to later in pregnancy because of the recognition that these higher doses were likely to result in severely impaired fertility, as observed in the NIOSH study. Whereas the NIOSH researchers chose to reduce the dosing level to address fertility problems caused at higher doses, the NTP study avoided the problem altogether by exposing the animals later in pregnancy to increase the likelihood of pup survival and thus researchers' ability to measure defects should they occur. By exposing the animals during organ development, the NTP study sought to increase the probability of seeing effects if they existed. What this decision ignored, however, was the possibility that the effects observed in the NIOSH study resulted from exposure early in development at lower dosing levels. The final article, published in the journal *Fundamental and Applied Toxicology*, reported no significant birth defects of the skeleton. However, the toxicity of the dosing levels was apparent, as a large number of the pregnant animals died.[63]

While both the reproductive and teratogenic studies pointed to the estrogenic activity of BPA, because of the high doses used the effects observed could not be disaggregated from effects of general toxicity. Therefore, little was learned about the effects of chronic, nontoxic exposures to estrogens, which had informed Hueper's and Smith's concerns in the 1950s and which had been the NIEHS's focus of research on DES and other environmental estrogens earlier in the 1980s. The study of BPA's estrogenicity was, therefore, largely absent from regulatory testing.

In 1988, the EPA used the NTP studies to issue the first safety standard for BPA. Since the EPA considered the chemical to be "noncarcinogenic" on the basis of the NTP's 1982 carcinogenesis study, the agency noted that its safety level could be "based on the assumption that thresholds exist for certain toxic effects."[64] In other words, this meant that a safety standard could be set below a toxic threshold level. In reviewing the available data, the agency selected what was at the time the most conservative dose found to have some toxic effect, as no level was found to have no adverse effect. They chose 1,000 ppm (or 1,000 mg/kg) BPA, used in the NTP carcinogenesis study.[65] In absolute terms, this dose was higher than those used in the reproductive toxicity test, but because the chemical was given to the animals through their food, while the reproductive test placed capsules under the animals' skin, the agency recalculated the dose on the basis of assumptions about the animals' food consumption per day. The agency then determined the lowest dose at which an observed toxic effect was observed—the "lowest observed adverse effect level," or LOAEL—to be 50 mg/kg/day. Next, to account for uncertainty in the estimates, and to account for the differences among species (extrapolating from animals to humans) and within species (individual sensitivity), the EPA divided this new LOAEL by 1,000, the so-called uncertainty factor. This resulted in a reference, or safety, dose of 0.05 mg/kg/day or 50 µg/kg/day.[66] BPA was declared safe at this level for the entire population.

The completion of several large studies on BPA's potential carcinogenicity, mutagenicity, and reproductive toxicity, and the establishment of a safety standard, in many ways fulfilled the vision outlined in the NIEHS mission statement. Here was a chemical that had been in use since the 1950s, with little to no information on its toxicity and no standard of safety. BPA was prioritized for review because of its high-volume production; it was put through a handful of regulatory toxicity tests, some of which had been newly developed; the available research was reviewed by the EPA and a safety standard set. The system for testing chemicals and managing risk appeared to have worked. But in the process, ambiguity about BPA's carcinogenesis was obscured by certainty of its noncarcinogenicity, and its estrogenic effects were minimized and largely overlooked. Whether the safety standard sufficiently protected public or worker health was a question unanswered by the high-dose studies. A standard for safety had been set at a level that had never been tested. Safety was, therefore, a presumption based on the logic of a threshold dose-response relationship and the assumption that reproductive effects observed would diminish with dose.

The dependence of regulatory tests on very high doses of exposure ignored any potential distinction between carcinogenicity and toxicity. Concerns raised by researchers such as Hueper and Smith had considered the fact that some chemicals such as estrogen appeared to be carcinogenic at nontoxic levels of exposure. The only difference between carcinogenicity and toxicity in the regulatory process was the presumption of the dose-response relationship: threshold for toxicity and nonthreshold for carcinogens. This simple distinction did not account for a variable—the timing of exposure—that was considered critically important in research on DES. The disconnect between the development of scientific knowledge about synthetic estrogens in the 1970s and 1980s and the questions asked about the risks and toxicity of BPA in the regulatory testing process revealed the serious limitations of and rigidity inherent in the formalized process of regulatory toxicology.

THE PRESENCE AND PARADOX OF ENVIRONMENTAL ESTROGENS

In 1982, the same year that the NTP released its report declaring that there was "no convincing evidence" of BPA's carcinogenicity, John McLachlan and his colleague Retha Newbold, both researchers at the NIEHS, published a laboratory-animal study that replicated the extensive reproductive tract abnormalities, precancerous lesions, and rare vaginal cancers reported in women exposed to DES while in their mother's womb. Their paper built on nearly a decade of research on the effects of DES exposure during fetal development at the NIEHS laboratory.[67] McLachlan's research team worked within the Transplacental Division of the Laboratory of Reproductive and Developmental Toxicology. *Transplacental* means they studied the effects on developing offspring of chemicals, from cigarette smoke to pesticides and DES, capable of crossing the placental barrier. McLachlan's research was founded on decades of research on DES and estrogen, including that of his mentor, Roy Hertz, chief endocrinologist at the NCI. Hertz had studied the carcinogenic effects of estrogens since the 1940s and, along with Hueper and Smith, had warned of the dangers of DES in the late 1950s.[68]

Beginning in the early 1970s, first as a postdoctoral fellow at the NIH and later at the NIEHS, McLachlan started to construct a model for understanding how a synthetic estrogen such as DES functioned at the molecular level and how this biological activity could alter reproductive development. One of his first papers on the topic established an essential piece of evidence: DES could cross the placental barrier.[69] Subsequent papers reported on a number of abnormalities, including precancerous and cancerous lesions and impaired reproductive capacity, in reproductive tract development in male and female animals due to exposures during fetal development.[70] In these transplacental exposure studies, McLachlan's team tested levels of DES well below the lowest level at which toxic effects such as weight loss were observed. For example, in the 1982 study, dosing levels ranged from 5 to 100 µg/kg (parts per billion, a thousandfold below the ppm or mg/kg used in the BPA carcinogenesis study). These were doses a thousand times or more less than the levels at which toxic effects of DES could be observed.

McLachlan's research on DES reproduced evidence from human studies that indicated that the timing of exposure was a critical variable in determining the observed adverse reproductive outcome. Epidemiological studies of DES, led by Arthur Herbst of Harvard University, who published the first study in 1971, subsequently found that the risk of cancer and precancerous conditions increased dramatically if a mother took DES early in pregnancy.[71] The effects of DES exposure appeared to be influenced by both the dose and the timing of the exposure during fetal development. If exposure occurred during critical periods when the reproductive organs were developing, even nontoxic levels of DES could restructure or reprogram tissue organization in ways that resulted in a number of abnormalities later in life.

Evidence that pointed to the importance of timing of exposure, though well recognized in developmental toxicity, presented a paradox to the dominant model for understanding and assessing carcinogenicity in the late 1970s and early 1980s. As with the study of BPA's carcinogenicity in the late 1970s, the regulatory protocol relied on the assumption that exposing animals to large doses of a potential carcinogen increased the likelihood of observing a carcinogenic effect, if one indeed existed, and that adult or prepubertal animals were acceptable models for evaluating risk. Such assumptions were further supported by the dominant theory of chemical carcinogenesis since the 1950s: the somatic mutation theory, which states that cancer results from out-of-control cell proliferation due to DNA mutation.

Indeed, supporters of the Delaney clause used this theory to support arguments for a zero tolerance for carcinogens, the contorted logic being that even one molecule of a carcinogen could result in cancer if it caused DNA mutation. However, ample evidence suggested cancer incidence rose among the elderly, which suggested the importance of cumulative exposures to carcinogens over a lifetime. In response to this understanding, researchers in the 1950s developed a multistage, or two-hit, model of carcinogenesis that contended that two (or more) mutating events are necessary to initiate disease. If cancers are more likely to arise due to multiple hits or exposures, then the chance of seeing a cancerous effect, if it exists, increases with chronic exposure. The NTP's carcinogenesis testing protocol, therefore, chronically exposed rodents throughout adult life.[72]

DES, which is a human and animal carcinogen, appeared to be an anomaly. Like estradiol (i.e., natural estrogen), DES was not mutagenic, but it did cause cancers in the children of women who took the drug during their pregnancies, particularly in those who took the drug early in fetal development. That DES is carcinogenic and nonmutagenic conflicted with the somatic mutation theory then dominating cancer research.[73] Estrogens had long proven to confuse cancer research. Does estrogen exposure protect against cancer or increase risk? If it is capable of both, what differentiates these effects?

Estrogen, as Dodds noted in the 1920s, has a powerful effect on the growth of cells, and cancer is a disease of out-of-control, abnormal cell proliferation. The study of hormonal carcinogenesis explored how this powerful biological activity can result in cancerous development; it differed from research in somatic mutation theory that focused on the isolation of genes that control cell proliferation, so-called oncogenes. Mutation in oncogenes can trigger abnormal cell proliferation and cancerous growth. Joan Fujimura, in *Crafting Science,* describes how theory and methodology worked in tandem (or converged) to produce knowledge about genetic contributions to cancer. For example, she details how the standardization of recombinant DNA, inbred mouse colonies, and mouse cell lines provided a static and controlled genetic environment that allowed scientists throughout the world to isolate and manipulate DNA in the search for genetic mutations that initiated cancerous tumors, thereby reconsolidating and reorganizing the theory that oncogenes were the key to understanding cancer.[74]

One significant contribution to integrating this theory with regulatory practice was the development of a screen for genetic mutation. In the mid-1970s, Bruce Ames, a biochemist at the University of California, Berkeley, developed a cheap and fast test of carcinogenicity that screened chemicals for their ability to cause genetic mutation in bacteria, the so-called Ames test.[75] The screen combined bacteria with human liver enzymes to determine whether the metabolite of a chemical compound caused genetic mutation. In the mid-1980s, Ames used the test to rank the potency of carcinogens—both naturally occurring and human-made ones. His work drew widespread public attention and criticism from environmental advocates, largely because Ames came to view concerns about synthetic carcinogens as overblown and as a contributing factor to excessive regulation. Ames's ranking system minimized the contribution of industrial carcinogens to cancer in humans—favorable news to chemical producers, which at the time faced increased public scrutiny and regulatory oversight. Ames became increasingly drawn to the ideological and political Right and he began to speak out in critical policy debates. Most notably, he opposed Proposition 65 in California, which passed in 1986 and states that no company shall knowingly expose a person to a chemical that causes cancer or reproductive toxicity.[76]

Ames's test provides a rapid and cheap means to identify mutagens, but it conflates carcinogenicity with genetic mutation and by definition ignores nonmutagenic pathways of cancer, notably hormones. By the early 1970s, the development of new understandings of the mechanisms by which hormones exert their effects at the cellular level infused new energy into waning interest in hormonal regulation of normal and cancerous tissue. A report prepared by the National Panel of Consultants on the Conquest of Cancer for the Senate Committee on Labor and Public Welfare in 1970, released in anticipation of the passage of the National Cancer Act in 1971, discussed estrogens at length as evidence of nongenetic causes of cancer and the reversibility of the disease. "Hormonal carcinogenesis is a good example of cancer induction by a disturbance of the body's regulatory mechanisms. Upon removal of the hormonal stimulus, the phenomenon of tumor progression may stop and the cells revert to less malignant states or even to latency." The "question of whether the cancerous change is due to an alteration of the hereditary material—*a genetic change*—or whether it is due to an alteration of a non-genetic biochemical process which somehow affects the hereditary material—*an epigenetic change*," the report noted, remained unresolved.[77]

Given evidence that tumors could be reversed and hormones could induce cancer, the report concluded: "There is no evidence so far that cancer is due to a mutation."[78] Research on estrogens pointed to possible epigenetic mechanisms of cancer development—broadly defined at the time as nongenetic biological processes. "There is ample experimental evidence both for the all-important regulatory role of hormones and for the hypothesis that continuous and excessive hormonal stimulation can so disrupt cell function as to induce cancer."[79] (It took another forty years before the President's Cancer Panel, in 2008–9, affirmed for the first time the role of hormonal alteration, epigenetics, and risks of early exposure for cancer development later in life. Despite the panel's prestige, the American Cancer Society dismissed the report's conclusion that environmental risks from chemical exposures had been woefully underestimated.)[80]

The suspected mechanism by which estrogen might induce cancer involved the hormone's powerful ability to trigger cell proliferation. Estrogen had long been considered the "ultimate *positive* signal" to cells to multiply.[81] Interest in estrogen's biological activity at the cellular level led Ana Soto and Carlos Sonnenschein, researchers at Tufts University Medical School, to the study of hormonal carcinogenesis and ultimately to environmental estrogens. In the early 1970s, Sonnenschein developed the world's first estrogen-sensitive cell line, a human cell that would proliferate in the presence of estrogen or estrogenic compounds. For his model, Sonnenschein used a human breast cancer cell line called MCF7, recently developed at the Michigan Cancer Foundation in Detroit. It took years to develop. The estrogen receptor gene and receptor protein, which binds the hormone and transports it into a cell nucleus, had to be isolated. Then Sonnenschein needed to develop a culture medium that would support cell life and would allow researchers to control the level of estrogen present.[82]

Prior to the development of a cell-based model, which Soto and Sonnenschein called the E-Screen, whole animals were used to test for the presence of estrogen. The basic procedure involved removing the ovaries—the principal source of estrogen—from female rodents, causing the uterus to atrophy and estrus to cease, and then administering estrogen or an estrogenlike compound to the animals. An increase in the size of the uterus, proliferation of the cells in the uterine lining, or the return of estrus signaled the positive presence of estrogen.[83] This simple depletion-repletion study—removing the ovaries as the main source of estrogen and then restoring the missing element and thus the original state of the animal—originally was used to identify and isolate estrogen in the late 1920s and early 1930s.[84]

Sonnenschein's cells allowed him to more narrowly focus on the biological activity of estrogen in the cell, independent of the whole body. While experimenting with the cells in the animal (in vivo), Sonnenschein reproduced the basic theory of estrogen activity. That is, when he injected the estrogen-sensitive cells into animals that had ovaries (and thus already had estrogen), the cells proliferated. However, when Sonnenschein attempted to reproduce this effect in vitro—that is, in cell culture, outside the whole organism—he observed a puzzling phenomenon. When the estrogen-sensitive cells were in culture without serum (part of the medium to help sustain the cells) or estrogen, they proliferated. Sonnenschein had expected the cells to be in a state of quiescence. This was a basic principle of cell biology. When he added the estrogen-free serum to the cells, they stopped dividing. And then, when estrogen was administered, the cells (once again as expected) proliferated. The paradoxical finding, Sonnenschein hypothesized, indicated that the original, default state of the cells appeared to be proliferation rather than quiescence. If this was the case, could the researchers still hold that estrogen was the "*ultimate* positive signal" for cells to proliferate, if they proliferated without estrogen? Adding to the confusion, his laboratory also observed that estrogen didn't cause proliferation in some normal tissue and cancerous tissue with estrogen receptors.[85] As researchers probed deeply into the molecular biology of estrogen, the hormone continued to confound assumptions and expectations about its functions.

Sonnenschein inverted a fundamental question in cell biology by asking not what caused quiescent cells to proliferate but what caused cells to be inhibited. Such a hypothesis ran absolutely counter to cancer research on oncogenes that focused on isolating genes that triggered cell proliferation as the cause of cancer. He used estrogen cells to test this hypothesis.[86] Soto, a medical researcher and molecular biologist from Argentina, joined Sonnenschein's laboratory in 1973. Admittedly "obsessed with the molecular revolution" at the time, Soto joined Sonnenschein in his pursuit, which led them to ask: If the default state of cells is proliferation, does this complicate or change the theory of cancer development? By asking this question, Soto recalled years later, she gradually moved away from a narrow or reductionist focus on the cell to consider more complex organizations of cells into tissues and organs. She queried: How do cells communicate with one another in biological development—what she and Sonnenschein later referred to as the "society of cells"—and what happens when this communication goes awry?[87]

Sonnenschein and Soto's estrogen research and intellectual boldness—whether ultimately accurate or not—ultimately led the Tufts team to the study of environmental estrogens and the work of McLachlan. While working on DES and other estrogenic compounds in the mid- to late 1970s, McLachlan drew from many disciplines, from toxicology and molecular biology to endocrinology. The downside to working across disciplines or, in Soto and Sonnenschein's case, against the dominant grain in cancer research meant that there was no shared research community of scientists studying estrogens or hormonal carcinogenesis. As part of a concerted effort to expand their lonely field and build a more cohesive thought community, McLachlan organized the first "Estrogens in the Environment" conference, in 1979.[88] The meeting aimed to gather, for the first time ever, leading scientists from federal institutes, agencies, and universities around the country who were working on various aspects of estrogen: the mechanistic understanding of estrogen, the structural diversity of chemicals with estrogenic activity, and research on the effects of exposure to estrogens, in particular DES. It was at this conference that Soto, Sonnenschein, and McLachlan met for the first time, and the Tufts researchers learned of synthetic estrogens in the environment, including chemicals in plastics and pesticides.

"Thousands of chemicals are introduced into our environment with little knowledge of their effects on two physiological processes which are central to our survival as a species—reproduction and development," McLachlan and NIEHS director Rall wrote in their opening remarks.[89] Estrogens in the environment included a host of compounds, including DES excreted by exposed livestock (researchers estimated that in 1971 over twenty-seven thousand kilograms of DES were given to livestock); industrial chemicals such as DDT and PCBs; and naturally occurring estrogens in plant life and foods, so-called phytoestrogens.[90] For decades, researchers such as Dodds in the 1930s and Bitman and Cecil in the 1960s had considered chemical structure to be a strong indicator of estrogenicity, but at McLachlan's conference researchers noted a diversity of chemical structures with reported estrogenic activity. Understanding how this diverse array of chemicals exerted estrogenic effects was a "black box" in molecular biology and a research domain ripe for serious scientific pursuit. Given the complexity of chemical structure, researchers questioned whether the mechanism of action for all estrogenic compounds was the same as that of natural estrogen, estradiol. The prevailing understanding of estrogen's mechanism at the time, as reviewed by Jack Gorski, father of estrogen endocrinology, involved the binding of estrogen to the estrogen receptor in the nucleus of the cell. The strength of an estrogenic effect, therefore, was determined by the affinity of the estrogen to the receptor, a protein molecule. Receptor-binding signaled to the cell to do something: proliferate, die (apoptosis), or allow other molecules into the cell.[91]

Six years later, when McLachlan convened the second "Estrogens in the Environment" meeting, its focus expanded to include the biological and toxicological effects of estrogenic compounds, with attention to studies of prenatal exposure, including work on DES. Because cell signaling is critical in directing the development of tissues and organs—what Soto and Sonnenschein later referred to as the "society of cells"—particularly in early development, researchers were beginning to consider the impacts of altering this communication with estrogenic compounds present in the environment. Wildlife researchers also attended the meeting and presented studies on the effects of DDT and PCBs on infertility, including low sperm counts and reproductive tract abnormalities.[92] The scientific community had strengthened and grown significantly. At a third, and even larger, international meeting, held in 1994, new techniques for detecting estrogen, new sites of estrogen biological activity such as the brain, the immune system, and neurobehavioral systems, and a new mechanistic understanding of estrogen's involvement in carcinogenesis dominated the meeting's proceedings.[93]

The emergent scientific research on xenoestrogens, or "estrogens in the environment," pointed to epigenetic alterations—nongenetic changes that affect an organism's development—rather than genetic mutation. It also raised important questions about the health risks of the millions of pounds of synthetic estrogenic compounds produced every year and increasingly detected in water, air, and consumer products, including the equipment used in research laboratories. Emerging evidence of the reproductive, developmental, and carcinogenic effects of synthetic and environmental estrogens seeded scientific debates that became overtly political when environmental advocates in the 1990s began to call for regulatory reforms.

The resounding lesson from DES research was that when it came to estrogens, determining carcinogenic effects was anything but straightforward. Effects differed significantly depending not only on the dose but on the timing of the dose as well. By the 1980s, molecular biologists studying how estrogens affect growth and development had opened up a complex field of research with considerable implications for how to test industrial chemicals with estrogenic properties. Sonnenschein and Soto investigated how estrogen signaling among cells directs the organization of tissues and organs and how disruption of that signaling by synthetic estrogens may result in increased susceptibility to cancer. Mechanistic understanding of estrogens, both natural and synthetic, expanded as well. Estrogens could bind, block, or partially bind to a number of estrogen receptors in various organs and tissues. Different effects could result depending on when exposure occurred in development, at which dose, and which system (e.g., the brain, the reproductive system, or the neurobehavioral system) was affected. The complexity of understanding the risks of exposure to a given estrogenic compound was becoming increasingly dense.

None of this emerging mechanistic understanding of estrogens was captured in the regulatory toxicity tests used to establish the first safety standard for BPA in 1988. While BPA's estrogenicity was recognized and noted by regulators, it was considered inconsequential to determining its safety. Compared with DES, BPA was shown to have weak estrogenic properties. While the EPA said it was as potent as the estrogenic form of DDT, they noted that it was not as persistent in the environment. BPA was declared safe at 50 µg/kg and was said to be "noncarcinogenic." As a consequence, any biological effects that might occur from estrogenic activities below the toxic threshold were presumed to present insignificant risks.

By the late 1980s, in the wake of increased federal investment in environmental health sciences and some expansion of federal testing requirements, two distinct pathways for producing knowledge about chemical risks had emerged: regulatory toxicity testing, conducted by private contracting laboratories for industry; and hypothesis-driven research, supported by federal grants, which continued to explore the biological effects of lower and lower concentrations of chemicals, including environmental estrogens. Communities of researchers developed to support both processes. McLachlan helped to build a community of largely academic scientists working across disciplines on environmental estrogens.

Some researchers with experience in regulatory toxicology found valuable positions in the private sector. For instance, the principal investigator of the NTP's BPA carcinogenesis study left his position at the private laboratory Litton Bionetics to work for Mobil Chemical Company. James Lamb, the director of the NIEHS's BPA reproductive toxicity test, left the institute for the private sector in the early 1990s, with a law degree in hand. With his deep knowledge of how the regulatory process worked on both the scientific and legal sides, Lamb took a position with the consulting firm Jellinek, Schwartz, & Connolly, Inc., which represented the interests of pesticide producers in navigating the regulatory process. Steven Jellinek, a senior partner in the firm, was the first administrator of the Office of Toxic Substances at the EPA in 1977.[94] Lamb later held the position of senior vice president of Applied Toxicology and Risk Assessment Practice at the Weinberg Group, a "product defense" consulting firm.[95]

Regulatory toxicity testing and hypothesis-driven research ran in parallel until the early 1990s, when their trajectories began to collide in science-policy debates over endocrine disruptors: chemicals with hormone-like properties, including environmental estrogens, that can adversely affect health. A small group of researchers studying environmental estrogens, among them Theo Colborn, Fred vom Saal, John McLachlan, Ana Soto, and Shanna Swan, began to more publicly raise their concerns not only about risks to human health but about the inadequacy of regulatory toxicology to study these compounds, testifying before Congress and appearing in documentaries on the topic. New studies on the biological effects of BPA at levels below the regulatory safety standard, conducted by vom Saal in the 1990s, challenged both the standard itself and the process by which the standard was set. The thesis of endocrine disruption ignited a fierce scientific and political debate in which BPA became a central molecular figure.

4 Endocrine Disruption

New Science, New Risks

In the early 1990s, scientists working in the Division of Endocrinology at Stanford University's School of Medicine stumbled upon an uncontrolled contaminant in their laboratory: BPA. It was scientific serendipity. What the researchers first thought might be an estrogenic substance produced by the yeast cells they were working with turned out to be BPA. Used to make polycarbonate plastic, BPA had leached out of the plastic flasks in the laboratory. Polycarbonate's strength, heat resistance, and clarity make it an excellent replacement for glass, and by the 1980s it was replacing glass tubes, flasks, and bottles in laboratories and hospitals. At the time when BPA "escaped" into the Stanford lab, polycarbonate production was booming. Annual U.S. production topped one and a half billion pounds in the early 1990s, and continued strong growth was expected for the foreseeable future.[1]

It didn't take the Stanford endocrinologists long to determine that the source of the estrogenic contaminant in their laboratory was the plastic flasks, and they published their findings in the journal *Endocrinology* in 1993.[2] A similar incident had occurred in a laboratory thousands of miles away a few years earlier. One morning in the late 1980s, Ana Soto and Carlos Sonnenschein of Tufts University arrived at their laboratory to find that the estrogen-sensitive breast cancer cells they had spent decades working with had proliferated without any known intervention (see chapter 3). After searching for months for the mysterious contaminant triggering cell division, Soto and Sonnenschein finally discovered that a chemical used in the plastic test tubes, nonylphenol, was leaching out and contaminating the cells. They eventually demonstrated that nonylphenol could bind to estrogen receptors and trigger the proliferation of the breast cancer cells. The Tufts team published this work in 1991.[3]

Serendipity continued to influence investigations of environmental estrogens and BPA. In 1998, Pat Hunt, a geneticist at Case Western University, noticed that nearly half of the female mice in her reproductive biology lab suddenly began displaying serious chromosomal abnormalities in their eggs. This was an obvious and dramatic increase from the normal background abnormality levels of 1 to 2 percent. She spent several years trying to assess what had contaminated her lab, and eventually determined that a temporary laboratory technician had inadvertently used a caustic industrial detergent to clean the polycarbonate cages used to house the mice.[4] Hunt hypothesized that the caustic cleaner broke down the polycarbonate, releasing BPA into the environment, resulting in chromosomal abnormalities in the exposed mice. To test her theory, she designed an experiment that purposely exposed mice to BPA. The study found significant increases in chromosomal abnormalities, confirming her original hypothesis.[5]

These accidental discoveries were significant moments of scientific and technological serendipity that made it possible to identify the presence of synthetic estrogens in the environment and to understand their effects on human biology. Progress in analytic chemistry, endocrinology, molecular biology, and genetics allowed these researchers to isolate smaller and smaller concentrations of chemicals contaminating their experiments and to measure subtle cellular level changes such as cancer cell proliferation and chromosomal alignment. These effects were not the dramatic toxic events—death, reproductive failure, or even cancerous tumors—commonly observed in toxicity testing. They were biological changes that might increase disease risk (e.g., cancer cell proliferation). This shift in effects measured had potentially revolutionary implications for the definition of safety and the scientific, political, and economic processes by which safety's meaning was informed.

These seemingly accidental discoveries of environmental estrogens received attention because they coincided with the emergence of the new thesis of endocrine disruption. In the 1993 issue of *Endocrinology* in which the Stanford researchers' finding on BPA appeared, Kenneth Korach from the National Institute of Environmental Health Sciences wrote an editorial that highlighted the "surprising" places estrogens seemed to be appearing in the environment.[6] Korach was one of the organizers of the first "Estrogens in the Environment" meetings, initiated by John McLachlan. This notion of estrogens as being widespread in the environment picked up on broader concerns of endocrine disruption that Theo Colborn, a zoologist, a pharmacist, and, at heart, a fierce environmental advocate, was beginning to articulate.

As this chapter narrates, research on environmental endocrine disruptors was thus thrust into political contestation. Not unlike climate change, endocrine disruption, at its broadest level, suggested that dramatic alterations in the environment caused by chemical pollution posed a significant threat to the long-term health, viability, and well-being of humans and wildlife. Advocates of the thesis perceived its biggest implications for public health to be the undermining of the human population's long-term reproductive capacity. Meanwhile, industry recognized that some chemicals elicited hormonelike effects but argued that their risks to human health were minimal.

As advocates of the thesis pushed for its legitimacy and acceptance throughout the 1990s, the scope and breadth of the meaning of endocrine disruption became highly contested at two levels (and at times advocates of the thesis had to answer both levels of criticism at once). The first level of the fight was over the theory of endocrine disruption itself. What was the scientific basis for its legitimacy? Was endocrine disruption simply a description of a chemical's biological activity (e.g., its ability to interact with estrogen receptors or disrupt thyroid production), or did it presume that such activity necessarily resulted in adverse health effects? If so, what defined these adverse effects? The most extreme critics decried the thesis as being grounded in "junk science" crafted by environmental advocates.

The second level of the fight was over regulatory response. When Congress mandated that the EPA develop a program for testing endocrine disruptors in 1996, conflicts erupted over the most basic task of defining an endocrine disruptor. The definition carried implications for whether a new approach to testing and setting regulatory safety was required to protect the public's health from this threat. Could low doses of endocrine disruptors present in the environment have large health effects, such as impaired human behavior, fertility, altered development, and increased risks of disease? Or was the human body resilient to subtle hormone changes?

The debate over endocrine disruptors in the late twentieth century raised the same fears about unchecked chemical pollution's long-term impact on human health and survival that had dominated the environmental movement since the 1960s. But as research continued, endocrine disruption presented a more nuanced challenge: how to integrate rapidly emerging research and proliferating data into decision making and chemical innovation.

A NEW TERM IS COINED

The phrase *endocrine disruption* was coined in 1991 at a multidisciplinary meeting of scientists convened by Theo Colborn and John Peterson Myers, president of a large philanthropic foundation, to examine the evidence on environmental estrogens' and other chemical contaminants' effect on the health of humans and wildlife. Colborn first had become aware of environmental estrogens while working on a comprehensive review of the health of the Great Lakes in the late 1980s. Through the course of her review, she observed what she believed to be a disturbing trend in the health of the region's wildlife. It appeared that a number of species were exhibiting reproductive and behavioral abnormalities in their young that threatened the survival of the population. The seemingly divergent effects all had one thing in common: they appeared to be manifestations of endocrine abnormalities. The story of her revelation has been told by Colborn herself in the popular book on endocrine disruption *Our Stolen Future*, and by Sheldon Krimsky in *Hormonal Chaos*. What is emphasized here is how personality and individual experience shaped the scope of the endocrine-disruption thesis and informed its political implications.

Colborn was an empiricist who used inductive reasoning to develop theories about complex and dynamic systems. She pulled information from research across the disciplines, reflecting the broad background of her unusual scientific training. For much of her adult life, Colborn lived and worked along the western range of the Colorado Rockies, where an irrigated desert landscape rises to meet majestic snow-capped mountains and energy companies mine coal and drill for natural gas. In these mountains, she learned science empirically. She practiced genetics as a successful sheep breeder and, like Rachel Carson, learned ecology as an avid birder and explorer of the natural environment around her. Working as a pharmacist in several area drugstores, Colborn also practiced chemistry. A keen observer of the individuals who visited the pharmacy, she became concerned by what she considered to be patterns of health problems exhibited by people living in similar watersheds. Curious about the impact of extensive mining on the mountain streams that fed the area's precious reservoirs, Colborn became involved in water-quality issues at the state level. After decades of work as a rancher and pharmacist, Colborn, at the age of fifty-one, decided to return to graduate school to study concentrations of trace heavy metals in the water and invertebrates of Colorado Rockies streams.[7]

In 1985, after seven years of graduate school, Colborn had a master's and a doctoral degree in hand and decided to accept a two-year fellowship at the Congressional Office of Technology Assessment (OTA) in Washington, D.C., where she would work on water and air pollution. Established during the Nixon administration, the OTA served as a think tank for Congress but was disbanded by the Republican-led Congress in 1995. Her work at the OTA plunged her into Washington politics, where she learned some of the ins and outs of Capitol Hill and the environmental nonprofit community. As her fellowship was drawing to a close, she was asked to join the staff of the Conservation Foundation, and in the late 1980s and early 1990s, working on behalf of the organization, she participated in an assessment of the environmental health of the Great Lakes led by the U.S.-Canadian International Joint Commission (IJC).[8]

The IJC was established under the Boundary Waters Treaty, signed in 1909 by the United States and the United Kingdom as a means to settle water quality, quantity, and access disputes between the United States and Canada. In 1972, the commission was charged with addressing the problems of transnational pollution as part of the Great Lakes Water Quality Agreement. The agreement brought increased attention to water pollution and called for expanded monitoring and assessment of the health of wildlife and humans in the Great Lakes region. Subsequent political interest and investment in scientific research perhaps produced more environmental data on ecological and human health in the Great Lakes area than in any other region in the United States or Canada. Growing evidence of harm to wildlife, fish, and humans caused by exposure to persistent toxic chemicals, including pesticides, dioxins, and PCBs, resulted in a renegotiation of the treaty in 1978 and the establishment of the ambitious goal to eliminate discharges of all persistent toxic chemicals into the lakes. A number of human studies funded by the United States and Canada examined the health effects of consuming Great Lakes fish, which were known to bioaccumulate persistent chemical pollutants, including PCBs. Alarmingly, epidemiological studies published between 1985 and 2000 found that women's consumption of Great Lakes fish before and during pregnancy reduced the intellectual performance of their children.[9]

The Conservation Foundation became involved with the IJC through the extensive political connections of the organization's top leadership, notably Terry Davies and William Reilly, in the late 1980s. Prior to coming to the Conservation Foundation, both had worked for Nixon's Council on Environmental Quality and were intimately involved in the original plans for structuring the EPA and in the crafting of the Toxic Substances Control Act. Reilly served as president of the Conservation Foundation in the mid-1980s, and Davies was executive vice president from 1976 to 1989. In 1985, Reilly moved from the Conservation Foundation to the World Wildlife Fund (WWF). He served as president until 1992, when he was appointed EPA administrator.[10] Colborn moved with Reilly to the WWF and remained there until establishing her own nonprofit, the Endocrine Disruptor Exchange, in 2002.

During the 1980s and 1990s, many large environmental advocacy organizations, including the WWF, shifted their focus away from the grassroots environmental activism that had dominated their 1970s work toward a new "middle ground" between ecological and economic concerns. The shift was part of environmentalism's significant turn away from the anticapitalist and antigrowth ideology of the late 1960s and 1970s toward a more moderate, business-friendly approach. In this new form, environmental organizations such as the Conservation Foundation and WWF strove to harness innovative technology and the tools of cost-benefit analysis to make a cleaner world and a stronger U.S. economy without reliance on command-and-control regulations. Mainstream environmental organizations (to the extent that they remained involved in issues of toxic chemicals—many scaled their programs down substantially) embraced the use of risk-assessment methods that had been the subject of debate between the EPA and OSHA in the 1970s.[11] For example, Davies served from 1989 to 1991 as assistant administrator for policy, planning, and evaluation at the EPA under George H. W. Bush but then returned to his earlier position at Resources for the Future, a Washington, D.C.–based think tank established in the early 1950s to develop economic analyses of natural resource use employing models of cost-benefit analysis.[12]

Colborn took a decidedly more radical approach to chemicals management than did the leaders of Conservation Foundation and the WWF. This was most apparent in her view of risk assessment, which she considered to be an insufficient and inherently flawed tool for evaluating chemical safety. Similar skepticism about risk assessment was shared by other, more radical environmental advocates, who distanced themselves from the EPA and mainstream environmental organizations, which, in shifting toward the political center, developed closer ties and relationships with large business interests.[13] From Colborn's perspective and in her experience, risk assessment effectively delayed decision making by binding up agencies in an endless cycle of analysis and review. Further, she contended that risk assessment frequently failed to result in decisions accounting for early-warning signals of degradation and harm in complex systems such as wildlife and human health.[14]

Despite her differences, Colborn's affiliation with the Conservation Foundation and the WWF provided her with the opportunity to serve as a member of the environmental health subcommittee of the Science Advisory Board of the IJC in the late 1980s. This work led her to the hypothesis that chemical pollution was undermining wildlife and human health through interaction with the endocrine system. In the course of her review, she read hundreds of wildlife studies conducted in the Great Lakes area that stretched back several decades. She also came to know several Canadian researchers on the commission—including Michael Gilbertson and Glen Fox, both of whom had studied wildlife health in the region for decades and were deeply concerned about the health of many species—and they strongly influenced her thinking. Gilbertson and Fox's work documented historical trends in wildlife health over the previous half century related to the rise and decline of persistent pollutants, such as PCBs and DDT.[15]

Colborn's Canadian colleagues introduced her to an important tool for assessing the weight of the evidence in the vast body of research she had accumulated.[16] In the broadest terms, "weight of the evidence" is a methodology used in law and science to evaluate the persuasiveness of data. In law, the strength of the evidence, or burden of proof, varies according to whether a case is a criminal or a civil one. In a criminal case the burden of proof for guilt is more demanding, requiring the weight of the evidence to be beyond a reasonable doubt, whereas in a civil case guilt rests on the preponderance of the evidence.

In science, however, models for assessing the weight of the evidence can vary considerably. Among the most frequently cited models for evaluating the evidence are Robert Koch's late nineteenth-century postulates for establishing a causal relationship in the study of infectious disease and Sir Bradford Hill's criteria for causality and correlation in the study of chronic disease. In a famous speech in 1965, Hill, a well-known British epidemiologist, outlined nine criteria for assessing the causal relationship between an exposure and a suspected effect in epidemiology, with particular attention to environmental hazards in the workplace. (The significance of Hill's criteria was that they addressed chronic disease at a time when infectious disease appeared to be on the decline in the First World because of public health interventions—e.g., clean water, vaccines, and rising standards of living.) Hill's criteria for assessing correlation and causality include the statistical strength of the association, consistency, specificity, temporality (exposure before outcome), plausibility, biological gradient (i.e., dose-response relationship), coherence, experiment, and analogy. In Hill's criteria, the rules are not hardened and can be adjusted over time in response to emergent data. As Hill remarked, despite his recognition that the criteria he laid out might be useful in studying association "before we cry causation," none were "hard-and-fast rules of evidence that *must* be obeyed before we accept cause and effect."[17]

In the 1990s, the Ecosystem Health Working Group of the IJC, which included Glen Fox, sought to adapt Hill's nine criteria to ecosystem biology and the research experience in the Great Lakes region, which they characterized as having multiple biological hazards functioning in a complex system and potentially associated with multicausal diseases and disorders that emerged over long latency periods. Given the considerable complexity of each factor—multiple hazards, ecological systems, and multifactorial diseases—that might influence the relationship between exposures and disease outcomes, demonstrating unequivocal certainty of causality would require a steep burden of proof. According to the IJC Working Group, using proof of causation as the standard for decision making meant that early-warning signals of declining ecological and/or human health might be missed. The challenge, then, was to develop a model for assessing the weight of the evidence that would allow for reliable early warnings and detection of health problems.[18]

In part, the group's approach attempted to correct an overreliance on statistical significance testing as the dominant indicator of strength of association. Hill had originally argued that because diseases were likely to be multifactorial, one should not, when evaluating the strength of an association, "dismiss a cause-and-effect hypothesis merely on the grounds that the observed association appears to be slight."[19] At the time Hill detailed his criteria, statistical testing was just becoming an essential and integral aspect of epidemiological study, and he warned of the dangers of a growing dependence on it to draw conclusions. "Far too often," he wrote, "we deduce 'no difference' from 'no significant difference.' Like fire, the X^2 test [chi-squared test] is an excellent servant and a bad master."[20]

The IJC adaptation of Hills's criteria was a response to the tendency of risk assessment to heavily rely on the statistical strength of an association between exposure and outcome. In exchange for greater statistical power, one accepts the higher statistical risk of stating there is no effect when there might be one (i.e., a type II error).[21] An overreliance on the statistical strength of association could mean that an early-warning signal is missed because the probability of concluding that there is no risk when a risk does, in fact, exist is increased. As Rachel Carson wrote in *Silent*

Spring decades earlier, "When one is concerned with the mysterious and wonderful functioning of the human body, cause and effect are seldom simple and easily demonstrated relationships. They may be widely separated both in space and time. To discover the agent of disease and death depends on a patient piecing together of many seemingly distinct and unrelated facts developed through a vast amount of research in widely separated fields."[22]

This "piecing together" is precisely what Fox and his colleagues, including Colborn, attempted to do. Drawing from ecological and social epidemiological models, the IJC Working Group integrated complex variables, such as multiple exposures, transgenerational exposures, bioaccumulation, and the long latency of disease, to present a fuller picture of where uncertainty existed and where there was evidence of serious effects. This was a deliberate effort to develop a process to assess the weight of the evidence that would allow for precautionary decision making in the face of uncertainty, a challenge posed not only by problems of ecological health but also by other complex scientific problems such as climate change.[23]

In the late 1980s and early 1990s, the precautionary principle appeared in a number of international treaties directed at sustainably managing global ecosystems, including the Montreal Protocol on Substances That Deplete the Ozone Layer in 1987 and the Third North Sea Conference in 1990.[24] The Rio Declaration on Environment and Development, a set of principles to guide global agreements that outlined sustainable development, signed at the first Earth Summit in 1992, included adherence to the precautionary principle, defined as follows: "Where there are threats of serious or irreversible damage, lack of full scientific certainty shall not be used as a reason for postponing cost-effective measures to prevent environmental degradation."[25] The codification of the precautionary principle in these treaties demonstrated an emergent political acceptance, at least in European nations, of the notion that in the face of ecological complexities, and given evidence of risks of irreversible environmental damage, some action or government intervention may be justified before a high burden of proof of causation can be met.

Colborn's weight-of-the-evidence evaluation of Great Lakes wildlife research sought to integrate the precautionary approach into the evaluation of complex evidence. It evaluated evidence in multiple species and drew extensively on historical data. The pattern of evidence observed suggested that the most serious health problems and abnormalities didn't appear in adults but disproportionately affected young offspring. The severest effects appeared in sixteen species with one common attribute: they were all at the top of their food chains. As top predators, these species, like bald eagles, had long been known to bioaccumulate many persistent chemicals, such as PCBs and numerous pesticides, in fatty tissue. Thus chemical concentrations increased at each level of the food chain because they persisted in the fat or tissue of prey. For example, an eagle eats fish that feed on smaller fish that eat smaller organisms or plankton, and all along the way chemicals stored in fatty tissue become more highly concentrated.

Because contamination is passed from the parental generation to offspring, particularly among top predators, adverse effects manifest more acutely in the younger generation. The concept of transgenerational effects became the central thesis in the IJC Working Group's report, published in 1990, entitled *Great Lakes, Great Legacy?*, which highlighted reproductive anomalies, immune problems, and behavioral abnormalities in the young. Tying all of these systemic health problems together, Colborn argued, was their relation to dysfunctions in the animals' endocrine or hormone system.[26]

After completing the report, Colborn found she couldn't shake the feeling that she had come across something with profound consequences for human health. She told herself that before she moved on to another project she would make one major push to have her thesis evaluated by the leading experts whose work had informed her report. Would these experts confirm her conclusions? Were some chemicals disrupting the functioning of the endocrine system? Were these chemicals affecting the developing young more than adults? Could the effects of exposure be passed from one generation to the next? And importantly, considering that many humans are top predators, what were the implications for human health? Looking back on that time, Colborn reflects that, considering what she knew, she felt it unconscionable to turn her back on the issue unless she was proven wrong.[27]

Colborn found her opportunity to test the Great Lakes thesis in the early 1990s, when she accepted a position as a senior fellow at the W. Alton Jones Foundation, under its new director, John Peterson (Pete) Myers. The Jones Foundation provided grants to organizations working on environmental sustainability and nuclear nonproliferation. Myers and Colborn had met several years earlier, in 1988, when Myers was vice president of science at the National Audubon Society. Trained as an ornithologist and acutely concerned about the impact of chemical exposure on migratory birds observed in his own field research, Myers immediately understood the serious implications of Colborn's writings on the Great Lakes. When Myers first accepted his position at the Jones Foundation, he negotiated a fellowship position for Colborn that provided her with the financial support and time necessary to focus exclusively on transgenerational effects of chemical exposure.[28] Myers also developed a small program at the foundation to support research and advocacy on the issue of chemical contamination, providing funding for research on estrogenic compounds, including support for John McLachlan's "Estrogens in the Environment" meetings.[29] Through Myers and the Jones Foundation, the threat of endocrine disruptors was introduced to environmental advocacy organizations, other foundations, and political champions of environmental issues who came to power during the Clinton administration, namely secretary of the Interior Bruce Babbitt and vice president and former senator Albert (Al) Gore (D-TN).[30]

Working together at the Jones Foundation, Colborn and Myers organized a multiday workshop, "Chemically-Induced Alterations in Sexual Development: The Wildlife/Human Connection," that brought together leading researchers in wildlife biology, estrogens, and endocrinology to evaluate the thesis that exposure to some chemicals in the environment might be undermining the health of wildlife and humans via disturbance of the hormone system. The meeting, held in 1991 at the Wingspread Conference Center in Racine, Wisconsin, included researchers such as Soto, Howard Bern, and John McLachlan, who had previously met at the "Estrogens in the Environment" meetings. Bern was a giant in the field of DES and had published some of the first articles on its developmental effects with Arthur Herbst (the scientist who published the 1971 epidemiological study reporting an increased risk of rare vaginal cancers in women exposed to DES in the womb).[31] Others in attendance had never before met or even known of one another's work. They included wildlife biologists who studied the health effects of chemical contamination, such as Lou Guillette, Michael Fry, and Glen Fox, as well as developmental physiologists, such as Fred vom Saal, who studied hormone regulation. Very few had met or knew of Colborn, even the major leaders in the field of estrogen research such as McLachlan and Bern. Recognizing that Bern's great stature as a leader in the field of estrogens and DES would add considerable gravitas to the meeting, Colborn was reassured by his agreement to attend but was also extremely nervous. When he entered the Wingspread Center, Colborn recalls, Bern exclaimed in a booming voice, "Where is this Dr. Theo Colborn? I want to meet her." Colborn, sitting nearby, rose to greet him. "I'm Theo," she said. Bern, who had just read through one of Colborn's articles on transgenerational health effects, declared, "I love you." The meeting was off to a good start.[32]

The structure of the meeting was unusual, but this was not a typical group of scientists. They all shared a confidence and interest in thinking and learning across their disciplines. This transdisciplinary engagement would come to mark the field of endocrine disruption for the next decade and beyond. The meeting itself involved several intensive days of sharing and discussing research, leading to a consensus statement on the state of the science on environmental estrogens and their impact on human and wildlife health.

The model for building a consensus statement came from recent work on climate change. The Intergovernmental Panel on Climate Change (IPCC), established in 1988, issued its first consensus statement on global climate change in 1990, which became the framework of the United Nations Framework Convention on Climate Change, the first global treaty aimed at reducing greenhouse gas emissions, signed in Rio de Janeiro at the Earth Summit in 1992. (The second assessment provided the framework for the Kyoto Protocol in 1997.) The IPCC organized the state of the science, as did the Wingspread participants, according to the following statements: "We are certain of the following . . . ," "We estimate with confidence that . . . ," "Current models predict that . . . ," and "There are many uncertainties in our predictions because . . ."[33] This process allowed a large group of experts to agree on what was known, what was likely to be true, and where they needed to focus future research. Looking for consensus allowed researchers to distinguish areas of greater and lesser certainty and uncertainty.

While uncertainty is inevitable and inherent in science, it often drives an endless political cycle of indecision: paralysis through analysis. Establishing consensus, therefore, provides a foundation for decision making in the face of considerable complexity and uncertainty, as in the case of climate change. In many respects, consensus building can support precautionary decision making by making evident what *is* known along with what isn't, avoiding the default position of suggesting that *all* the evidence must be in—an impossible task—before *any* possible preventative action can be taken. For example, from its initial report on climate change, the IPCC's certainty that the weight of the evidence indicates global warming is attributable to human activities has grown stronger with each subsequent report. Efforts to undermine the strength, validity, and rigor of this certainty persist today, and the IPCC reached a low point in 2009 when false accusations of scientific fraud were thrown at leading climate scientists: the so-called Climate-Gate. While the debacle produced no serious scientific challenges to the overwhelming consensus on global warming, the political effect was very real. Fewer Americans now believe that climate change is a real phenomenon.[34]

Pete Myers, who had closely observed the consensus-building process on climate change, introduced the model as the framework for the Wingspread meeting. He also coined the phrase *endocrine disruption* in the months leading up to the meeting. While mulling over a possible title for a climate change paper he was developing for the National Audubon Society, Myers had come up with the phrase *climate disruption.* As opposed to *climate change* or *global warming,* which he found too benign, *disruption* conveyed the message that manmade carbon emissions are wreaking havoc on the climate, with severe implications for the ecosystems upon which life depends. The term was edgy and politically provocative. In considering extant research on wildlife, studies of DES in humans, and experimental laboratory studies, Myers believed that the potential predicament was not simply a matter of endocrine change but a disruption to normal development and functioning that compromised health.[35]

At the end of three long days of meetings at the Wingspread Conference Center, the participants produced a consensus statement entitled "Chemically-Induced Alterations in Sexual Development: The Wildlife/Human Connection." The statement declared with certainty that "a large number of man-made chemicals that have been released into the environment, as well as a few natural ones, have the potential to disrupt the endocrine system of animals, including humans." The most profound effects, the scientists concluded, result from exposure during early (i.e., fetal and neonatal) development. Effects extend beyond cancer to include reproductive, immunological, behavioral, and neurological abnormalities and diseases. The consensus statement listed several dozen chemicals "known to disrupt the endocrine system," including DDT, PCBs, dioxins, kepone, lead, cadmium, and mercury.[36]

In its simplest iteration, the endocrine-disruption thesis holds that a number of synthetic and naturally occurring chemicals that mimic, block, or alter hormone function (not only of estrogen but also of androgen and other hormones) can alter the healthy development and functioning of the endocrine system. These alterations may manifest in a spectrum of adverse health effects and abnormalities in systems regulated by the hormonal system, particularly when exposure occurs during early development.[37]

The hormonal signaling system is a finely tuned chemical communication network of organs and glands that secrete hormone molecules into the bloodstream. Hormones carry chemical messages by binding to receptors in glands, tissues, and the brain. This elaborate system transmits information governing growth, development, mood, reproductive development (e.g., puberty), and metabolism (e.g., insulin regulation and temperature). Hormones are produced in accordance with a precise feedback system, directed by the brain, that maintains homeostasis, or dynamic equilibrium, around set points in the body that are exquisitely sensitive to alterations in the environment, circadian rhythm, and the nervous system.[38] If this system is disrupted very early in development, the Wingspread participants argued, the effects may not at first be noticeable, but as the organism develops, a small organizational change may manifest in significant changes or abnormalities in function and health.

For Colborn, the consensus statement provided strong confirmation of the concerns expressed in the Great Lakes report. The next step was to bring the statement to the public and raise the alarm about this issue—to draw public and political support, attention, and resources to this overlooked health threat. She jumped into action, writing articles and eventually a book, testifying on Capitol Hill, and meeting with legislators, reporters, regulators, scientists, and advocates. Just a few months after the Wingspread meeting, Colborn found herself on a plane with Reilly, who was then administrator of the EPA. Both had just attended an IJC meeting in Canada, where Reilly had given the keynote speech, and were headed back to Washington, D.C. Colborn used the opportunity to update Reilly on her work, handing him a newly faxed copy of the consensus report and summarizing its major conclusions. The next day, upon Reilly's request, she faxed copies of the statement to his office at the EPA.[39]

Colborn, together with her new colleagues from the Wingspread meeting, sought to disseminate their thesis into the scientific community to stimulate much-needed research interest. The first major endeavor was her edited textbook on the topic, *Chemically-Induced Alterations in Sexual and Functional Development*, published in 1992. Howard Bern wrote the opening essay, "The Fragile Fetus," which discusses how exquisitely sensitive or susceptible the developing fetus (not just the human fetus, but those of all animals) is to endocrine-disrupting chemicals. The fetus is fragile, he explains, because during this period of development, critical systems—reproductive, immunological, neurological—are structured and organized as cells proliferate and die. Endocrine-disrupting chemicals, Bern argues, may alter or disrupt the normal or healthy development of these critical systems, but the impact of such alterations or disruptions as disease or abnormalities may not manifest until the organism or individual has developed over time.[40]

One of the first overview articles on endocrine disruption to appear in a mainstream scientific journal was published in *Environmental Health Perspectives*, the journal of the National Institute of Environmental Health Sciences, in 1993. What is significant about this article, coauthored by Colborn, is that it succinctly outlines the meaning of endocrine disruption and, in so doing, explicitly points to implications for regulatory decision making. Endocrine disruption, according to Colborn, vom Saal, and Soto, indicates that, first, effects of exposure can be transgenerational (or transplacental) and therefore manifest in offspring; second, the effects of exposure are dependent on the timing of exposure during early development, not only on the amount of exposure (i.e., timing, rather than simply dose, determines the effect); and, third, effects may appear only as the offspring develops throughout life (i.e., such effects often are not present at birth). Endocrine disruption, they emphasize, is more than a mechanistic action—binding to or blocking a hormone receptor. Endocrine-disrupting chemicals present potential risks to multiple systems in developing organisms, even at the very low levels found in the mother's tissue, blood, and milk. These considerations of timing of exposure, low doses of exposure, and endocrine-dependent developmental health effects (e.g., reproductive abnormalities, behavioral problems) formed the early structure of the endocrine-disruption thesis.[41]

These distinguishing characteristics, however, were overlooked in regulatory toxicity testing. Endocrine disruption, therefore, was evidence of regulatory failure and of the need for reform of risk assessment and testing. Indeed, on many occasions Colborn called for the overhaul of the regulatory system along with increased funding for research on endocrine disruption. Such calls for reform and investment found a more receptive audience with the arrival of a new administration in Washington. In 1993, William Clinton was sworn in as president, ending twelve years of Republican leadership. Democrats also controlled Congress. For environmental organizations, the political moment was ripe for reform.

CHEMICAL REFORM IN THE CLINTON ERA

Clinton did not come to Washington with a stellar environmental record. For environmental advocates, however, his selection of Senator Al Gore as his vice president signaled that his administration would make clean air, water, and food a priority. As a senator, Gore had a solid environmental voting record and authored the environmental polemic *Earth in the Balance* (1992). Clinton made other key appointments that leading environmental organizations strongly supported, including Bruce Babbitt (former president of the League of Conservation Voters and governor of Arizona) as secretary of the Interior, Carol Browner (who previously had worked as an aide for Gore and directed the Florida Department of Natural Resources) as administrator of the EPA, Lynn Goldman as assistant administrator of the Office of Prevention, Pesticides and Toxic Substances at the EPA, and Hazel O'Leary (who early in the administration supported the release of secret government radiation experiments) as secretary of Energy.[42]

When Clinton came into office, pesticide reform was at the top of the agenda for environmental advocacy organizations, including the Natural Resources Defense Council (NRDC) and the newly formed Environmental Working Group, as well as for agricultural, food, and chemical industries. Both industry groups and environmental organizations pressed the Clinton administration to remedy the considerable confusion within federal pesticide regulations for food. The confusion stemmed from the fact that pesticide residues were regulated under two separate laws that promulgated distinct standards. First, the Federal Insecticide, Fungicide and Rodenticide Act, passed in 1947, regulated pesticide use on raw foods. (In 1972, the law was reformed to include the requirement that safety standards be based on cost-benefit analysis.) Second, the Federal Food, Drug and Cosmetics Act of 1958 regulated pesticide tolerances on processed foods and included the controversial Delaney clause for carcinogens. The chemical industry's opposition to the Delaney clause was clear from the very beginning, but by the late 1980s and early 1990s, mainstream environmental advocates had begun to concede that the clause failed to reasonably protect public health when it came to pesticides. This was due in part to regulatory confusion: even if the Delaney clause was applied to pesticides on processed foods, there still remained poor safety standards for residues on raw foods.[43]

These two different standards generated legal and regulatory confusion within the EPA.[44] In 1985, the National Research Council (NRC) of the National Academy of Sciences (NAS) released a report that recommended setting single pesticide residue standards for food (raw and processed) that would include standards for carcinogens. At the time, the EPA adopted these recommendations as if they were a legislative directive and began setting safety standards for carcinogenic pesticides.[45] Arguing that this violated the Delaney clause, the State of California, the NRDC, Public Citizen, and the AFL-CIO brought suit against the EPA in 1989. The settlement of the suit in 1994 upheld the Delaney clause and ordered the EPA to remove thirty-six carcinogenic pesticides from the market. The court decision provided a strong legal incentive for industry to agree to some reforms, and by the mid-1990s Congress began holding hearings on pesticide regulations.[46]

For researchers and advocates concerned about the threats of endocrine disruptors and the failure of the regulatory system to adequately account for their risks, pesticide reform legislation offered a unique window of opportunity to broaden the discussion of chemical risks. In a 1994 Emmy Award–winning documentary on endocrine disruption entitled *The Estrogenic Effect: Assault on the Male*, Colborn urgently calls for broad, non-specific reform of chemical regulation. "We have to . . . revisit every piece of legislation that is coming up for reauthorization to make sure that we include not only cancer as a risk element but . . . transgenerational health effects, the effects on the developing endocrine, immune, and nervous systems, which are all linked."[47] Colborn sought to have the film shown to as many politically influential people in Washington as possible. She traveled to Capitol Hill on a number of occasions to show it to groups of lawmakers and, along with her colleagues Soto and McLachlan, testified before congressional committees on the thesis of endocrine disruption.[48] She took the film to the EPA to show it to Carol Browner and her staff, and she lent her copy to the editor of the *Washington Post* and his wife, dropping it off at their Georgetown apartment.[49]

Advocacy groups around the country were also organizing around the issue of endocrine disruption. The Long Island–based breast cancer advocacy group One in Nine brought the issue to their representative in Congress, Senator Alfonso D'Amato (R-NY). Born in Brooklyn and raised on Long Island, D'Amato drew considerable political support from the Republican Party in suburban Nassau County. Loyalty to his voters, rather than ideological affiliation with the Republican Party, however, directed D'Amato's actions and priorities. Voter concern about the health effects of pesticides was a nonpartisan issue. When his constituents from One in Nine came to him with concerns about the relationship between pesticides and breast cancer, D'Amato helped secure government funding for the Long Island Breast Cancer Study, a major project initiated in 1993. Further, urged by the advocacy group and others, notably the NRDC, D'Amato attached to the pesticide reform legislation language requiring the EPA to test for estrogenic compounds.[50]

While the chemical industry welcomed reform of pesticide regulation, it generally considered additional endocrine-disruption testing requirements an unnecessary and costly burden and sought to block such provisions. In the early to mid-1990s, the leading industry trade associations worked together through the Endocrine Issues Coalition, which, according to an internal memo by the group, coordinated industry response to the "endocrine issue," facilitated the "harmonization of the industry position," "resolve[d] scientific uncertainties," and coordinated research. The coalition comprised the key trade associations—the Chemical Manufacturers Association (formerly the Manufacturing Chemists' Association); the Society of the Plastics Industry (SPI), which included the Bisphenol A Group; the American Crop Protection Association; the American Forest and Paper Association; and the American Petroleum Institute—and met quarterly during the pesticide reform debates.[51]

To counter Colborn's presence on Capitol Hill, representatives of Dow Chemical Company circulated its "Position on Endocrine Disruptors" to congressional offices in 1994. The Dow memo cast considerable doubt on what was very specifically and dismissively referred to as "Colborn's hypothesis." While it suggested that more research would be needed before any new testing requirements were made, it also sought to dismiss the urgency of the issue. Dow contended, first, that existing regulatory toxicology tests required by pesticide regulations were sufficient for evaluating endocrine-disrupting chemicals; second, that the weak estrogenicity of "even the most potent endocrine disruptors" suggested that these chemicals presented minimal risks; third, that the most potent chemicals had been banned; and, fourth, that natural estrogens as well as synthetic ones existed in the environment.[52] In other words, endocrine disruption was not a unique threat and therefore did not demand any reforms in chemical testing or regulation.

The opportunity for legislative reform to address endocrine disruptors came after the midterm elections of 1994, when the Republican Party brought an end to Democratic control of both houses of Congress and forty years of Democratic-majority rule in the House. With Republican power riding high on the wave of Congressman Newt Gingrich's popular Republican Revolution, chemical and plastics industry representatives and lawyers welcomed a more favorable climate for achieving long-sought reforms, namely harmonization of the pesticide laws and development of a more rapid program for reviewing indirect food additives. Jerome Heckman, the leading lawyer for the SPI, recalled that after the midterm elections Republican lawmakers brought industry leaders together to ask them which reforms they would like to see moved through Congress.[53] High on the priority list for the food, chemical, and agricultural industries was reform of the 1958 food law, and specifically relief from the 1994 court ruling that upheld the EPA's authority to ban a number of carcinogenic pesticides under the Delaney clause.

As Congress debated pesticide reforms in the mid-1990s, the issue of endocrine disruptors gathered political momentum. Three more consensus statements on endocrine disruption were produced in 1995 from two subsequent gatherings at Wingspread (II and III) and one in Erice, Italy.[54] The EPA, the CDC, and the Department of Interior under Babbitt requested that the National Academy of Sciences' (NAS's) National Research Council (NRC) form an expert panel to study the issue, a process that would take several years to complete. Anticipation of a popular book on endocrine disruption, *Our Stolen Future*, written by Colborn, Myers, and an environmental journalist, further contributed to rising political pressure on Congress to take some action in response to this threat.[55] Written for a lay audience, *Our Stolen Future* outlines the potential threats of endocrine disruptors to the human species, pointing to global declines in sperm counts and increases in female reproductive disorders. Release of the book, which included a foreword by Vice President Gore, came just months before Congress passed pesticide reforms in 1996.

After months of mediated negotiations between environmental advocates and the pesticides industry, a compromise was reached, and in August Congress unanimously passed the pesticide reform law, the Food Quality Protection Act (FQPA). The law created a single safety standard for all pesticides in food—raw and processed—and lowered tolerance limits to a determination of a one-in-a-million cancer risk. Safety standards would be risk based, ending the forty-year-old Delaney clause for pesticides. The intent of the one-in-a-million cancer risk standard was to protect infants and children by lowering many standards. Although there was a split within the environmental community over the loss of the Delaney clause, many mainstream organizations applauded the new law for its lower, single safety standard, its right-to-know provisions for consumers, and its requirement that the EPA review all existing safety standards on pesticides. The law was a significant victory for the pesticide industry because it simplified the regulatory process and removed the threat of a total ban under the Delaney clause as directed in the 1994 court ruling. This kind of compromise between the environmental community and industry reflected the Clinton administration's emphasis on finding solutions with "economy-ecology compatibility."[56]

For Heckman, the moment for his long-sought reforms of the 1958 Food, Drug and Cosmetics Act for indirect food additives had finally arrived. In 1997, Congress passed the Food and Drug Administration Modernization

Act, which established a new food contact notification program, devised by Heckman to speed up the clearance of food additives (renamed food contact-substances) by setting a 120-day deadline for the FDA to respond to notifications of new substances. Notifications to the agency include the identity of the additive, its use, and any available information demonstrating the safety of its use. The toxicological information required depends on the estimated dietary exposure to the substance. The lower the exposure, the less data are required. For instance, for exposures of less than 0.5 ppb, no data are required; for exposure within the range of 0.5 to 50 ppb, studies of genotoxicity (DNA mutation) must be conducted to indicate carcinogenicity risk.[57] The reformed law clearly tied data demands to exposure with the assumption that at very low levels only marginal risks exist and that consequently minimal or no testing is required. No discussion of endocrine-disruption testing was included in these reforms, as advocates focused on pesticide legislation. This was the very system that Heckman had sought to introduce in 1958, and, indeed, he had outlined the parts of the new law that affected indirect food additives.[58] For Heckman, the program sought to correct an egregious flaw in the original law. As he explained: "I have spent my professional life carrying on a crusade against the gross overregulation of food packaging materials." Drawing on biblical metaphors (not often seen in the pages of the industry-supported scientific journal *Regulatory Toxicology and Pharmacology*, on whose editorial board Heckman held a seat at the time), he expressed his unequivocal disdain for the 1958 law as it related to indirect additives: the "genesis," he wrote, of this "original sin" (that is, the 1958 Food, Drug and Cosmetics Act Amendments) gave rise to the "evil" conception of the "almost oxymoronic phrase" *indirect food additives* (chemicals not deliberately added to food, such as chemicals migrating from plastic packaging).[59]

Tucked into the FQPA and the reauthorization of the Safe Drinking Water Act (SDWA), which passed on the heels of the new pesticide law (also with overwhelming support from industry and environmental representatives), was language from D'Amato's bills that directed the EPA to regulate all estrogenic chemicals (not exclusively pesticides) in food and water supplies. With input from officials at the EPA, the original language of D'Amato's amendment was expanded beyond estrogenic chemicals to include testing for other endocrine effects.[60] The final version of the FQPA and SDWA directed the EPA to "develop a screening program, using appropriate validated test systems and other scientifically relevant information, to determine whether certain substances may have an effect in humans that is similar to an effect produced by a naturally occurring estrogen or other such endocrine effect as the Administrator may designate."[61] Congress set an ambitious time frame for the EPA to complete this task. It was instructed to develop the program by 1998, implement it by 1999, and report to Congress on its progress by August 2000.[62]

The EPA faced a formidable challenge. It was tasked with defining the scope of a testing program for chemicals based on a scientific thesis that was hotly contested. With the release of *Our Stolen Future*, the topic of endocrine disruption gained greater public attention, but it also quickly moved the thesis from the margins of scientific inquiry into the highly politicized space of public opinion. Were endocrine-disrupting chemicals threatening human health? Was toxicity testing capable of defining their risks? Or was the thesis an environmental polemic based on junk science?

ENDOCRINE DISRUPTION AS JUNK SCIENCE

Our Stolen Future tells the story of how the thesis of endocrine disruption emerged from disparate data. Much like *Silent Spring*, the book sounds an alarm about the effects of unchecked chemical exposures on human and wildlife health. The authors weave together the story of DES, wildlife, and human research in the Great Lakes that Colborn gathered in the late 1980s, as well as the research of Soto and Sonnenschein on breast cancer cells and the estrogenicity of nonylphenol. They recount Fred vom Saal's research on fetal exposure to various levels of hormones and the controversial data on declining sperm counts around the world; they introduce readers to the endocrine system and explain how persistent chemicals bioaccumulate up the food chain. The book is a polemical account of a scientific thesis. Much to the chagrin of the chemical industry, the authors make strong claims that estrogenic and other hormone-disrupting chemicals found throughout the environment are capable of undermining human fertility, reproductive health, intelligence, and behavior at the population level. Such concerns go beyond cancer, fear of which long informed the environmental movement of the 1960s and 1970s; endocrine disruptors, they argue, threaten the survival of humans and wildlife. Directed at a popular audience, the book is a call to arms.

According to one pesticide company representative, the book shook the industry to its core. "This is not just another Alar scare. . . . This will be trench warfare," one representative allegedly remarked at a briefing of the American Crop Protection Association, which represents pesticide producers.[63] In response to "media calls" and coverage of the book, the Chemical Manufacturers Association developed a consistent message that

Figure 4. *Our Stolen Future* authors at the edge of the Black Canyon in Colorado, 2006. *Left to right:* J.P. Myers, Dianne Dumanoski, and Theo Colborn. (Photo credit: J.P. Myers.)

highlighted uncertainty about the effects of endocrine disruption. The association distributed a list of vetted researchers and experts on the topic to help shape its message.[64] The overarching strategy was to question the legitimacy of the theory of endocrine disruption, point to the considerable uncertainty in the serious claims made in the book, and cast the book and the theory, together, as junk science.

Discrediting endocrine disruption as junk science was a strategy borrowed from the tobacco and chemical industries' well-used playbook. Although there existed substantial uncertainty about the scope and magnitude of the problem that endocrine-disrupting compounds presented to human and wildlife health, discrediting all of the work as junk directly sought to delegitimize all aspects of the thesis. As a tobacco executive explained in 1967, "Doubt is our product since it is the best means of competing with the 'body of fact' that exists in the mind of the general public."[65] David Michaels's book on the subject, *Doubt Is Their Product*, documents a number of disinformation campaigns coordinated by tobacco and chemical trade associations, public relations firms, and product defense firms. Naomi Oreskes and Erik M. Conway, in *Merchants of Doubt*, detail how a loose network of scientists and science consultants have effectively manufactured doubt about the risks of tobacco smoke, coal-burning power plants, and human impacts on climate change.[66] Doubt can be cast and propagated by highlighting uncertainty in research and questioning the legitimacy of damaging research. As long as doubt persists, products can stay on the market and restrictive regulations can be delayed.

The proliferation of junk science, according to the conservative scholar Peter Huber at the Manhattan Institute, could be attributed to the growing number of toxic tort cases flooding the courts beginning in the 1970s, involving, in particular, asbestos and lead. Indeed, the onslaught of large settlements in the United States by the late 1980s nearly bankrupted the global insurance giant Lloyds of London.[67] In his book *Galileo's Revenge*, published in 1991, Huber coined the term *junk science*.[68] The phrase quickly permeated into discourse on tort reform—a long-term objective of the chemical industry and tobacco industry. (In the mid-1970s, with insurance premiums skyrocketing because of problems related to "fire, toxicity, and product performance," the head of the plastics trade association coordinated an industrywide tort reform strategy.)[69] In the case of lead, research documenting neurological impacts in children exposed to low levels of lead presented a serious crisis for the industry already facing a deluge of liability claims. In the mid-1980s and 1990s, the lead industry launched a series of personal attacks on one of the leading researchers on low-level lead effects, Herbert Needleman, charging him with scientific fraud. He ultimately was acquitted of the charges, but only after an extensive investigation and persistent attacks on his professional integrity.[70]

Casting doubt on research by calling it junk science, combined with ad hominem attacks on individual researchers, is, at its most basic level, a strategy to keep damaging research out of the courtroom. This is not to suggest that the issue was so starkly black and white and that no abuse of science in tort cases was happening. It takes only a few unprincipled liability lawyers to fuel demands by industry leaders for processes to better manage the quality of science used in the courts and by other decision makers, such as regulatory agencies. The response to the alleged use of junk science in tort cases was to define what was considered to be legitimate or "sound science."

In 1993, the Supreme Court, in *Daubert v. Merrell Dow Pharmaceuticals, Inc.*, handed down a ruling on a toxic tort case that changed how scientific evidence would be evaluated in the courtroom. The petitioners in *Daubert* were children born with birth defects that the plaintiffs alleged had resulted from their mothers' ingestion of the drug Bendectin during pregnancy. At issue was the use of science to demonstrate a relationship between the drug and the birth defects. The plaintiffs founded their case on laboratory-based evidence and an unpublished reanalysis of epidemiological studies. The Supreme Court's majority decision overturned the common-law practice outlined in *Frye v. United States* (1923) that deferred to the scientific community to determine expertise. The ruling under *Daubert* made judges of the lower courts the "gatekeepers" of scientific evidence, with responsibility for distinguishing junk science from sound science.[71] In the case of Bendectin, the lower court had ruled in favor of the defense, finding that the scientific evidence brought by the plaintiffs did not demonstrate a causal relationship as accepted in epidemiological practice. In an amicus brief, the U.S. Chamber of Commerce, the Washington Legal Foundation, the National Association of Manufacturers, the Business Roundtable, and the Chemical Manufacturers Association supported the court's decision as a means to control the escalating problem of junk science.[72]

Strategies to undermine junk science and uphold sound science focused on secondhand tobacco smoke, air pollution, toxics, and climate change.[73] The development of organizations committed to promoting sound science and disclosing junk science through the media brought the issue more directly into the public domain. The Advancement of Sound Science Coalition (TASSC), a nonprofit organization established in 1993 by Phillip Morris and the public relations firm APCO Associates, coordinated efforts to challenge legislative actions on environmental tobacco smoke. TASSC promoted the language of sound science and the belief that junk science too often resulted in poor public policies and unnecessary expense—not only through tort litigation but also as a result of unnecessary regulation.[74] Now defunct, in the 1990s TASSC developed an extensive network of scientists, legislators, and local politicians dedicated to debunking evidence of the harmful effects of environmental tobacco smoke. The organization immediately expanded to cover the hot-button topics of global warming and endocrine disruption. In a memo to its supporters, TASSC's activities in 1995 were listed as including an "outreach program to mobilize support for regulatory reform," "collecting sound science 'horror' and 'success' stories," and "coordinating production and placement of a series of opinion-editorials on behalf of selected TASSC scientists, state legislators and local officials."[75]

In 1997, with continued financial support from Altria, the parent company of Phillip Morris, and ExxonMobil, TASSC came under the new direction of Steve Milloy.[76] A lawyer by training and an adjunct scholar at the libertarian think tank Cato Institute, Milloy set up the website junkscience.com and established the Free Enterprise Action Institute, both based from his home in Potomac, Maryland. A frequent commentator for Foxnews.com, he spoke out against evidence of global warming, the adverse health effects of secondhand smoke, and endocrine disruption. Milloy derisively suggested the theory of endocrine disruption was "closer to junk than 'junk science.'" Scientists publishing this work were quacks, he said. Milloy labeled Fred vom Saal, a participant in the Wingspread Conference in 1991 and soon to be an outspoken BPA researcher, the "cult leader."[77] In 1996, the year junkscience.com went online, Milloy turned his sights on *Our Stolen Future* and its supposed propagation of junk. In "response to unfounded environmental doomsday predictions in the book," Milloy formed a "truth squad" as part of TASSC's "Facts Not Fear" campaign.[78]

Milloy was joined in his attack by the fiercely pro–big business, antienvironmentalist Elizabeth Whelan of the American Council on Science and Health. Whelan originally founded this organization in 1977 in response to "emotionally-laden, exaggerated, non-scientific charges, which may lead to the banning or restriction of perfectly useful, and in many cases, health-promoting, substances."[79] She has defended the safety of Agent Orange, DDT, lead, asbestos, and various food additives. Whelan's position pleased many leading chemical and petrochemical producers, such as Dow, Exxon, Union Carbide, and Monsanto, which subsequently funded her efforts.[80] The plastics industry began supporting Whelan in the mid-1970s after she wrote her first book, *Panic in the Pantry*. The book challenges growing public concerns about the dangers of pesticide residues. The plastics industry was encouraged by Whelan's "common sense" in light of the "shenanigans" of the FDA, which at the time was attempting to restrict several food additives, such as acrylonitrile and PVC, under the Delaney clause.[81]

For Whelan, endocrine disruption represented the latest in a string of emotionally driven scare stories perpetuated by the environmental community, which grossly exaggerated scientific evidence. She published a long commentary on *Our Stolen Future* that aimed to dispel several of its central arguments by focusing on the presence of natural estrogens in the environment and the purported lack of evidence for a relationship between chemicals and sperm count declines or a relationship between chemicals and cancer.[82]

Skeptics of endocrine disruption felt confirmed in their beliefs when a major paper from John McLachlan's laboratory had to be withdrawn in 1997. McLachlan's research lab had published a paper in *Science* that reported a hundred- to thousandfold increase in the potency of weak estrogens when examined in mixtures. After failed attempts to replicate the findings, McLachlan publicly retracted the study. As it turned out, a postdoctoral student working on the study had forged the data.[83] While handled appropriately and quickly, the incident fueled arguments that endocrine disruption was junk science. Despite subsequent replication of the additive and synergistic effects of estrogens by many different laboratories, the retraction of this one study continues to cast a shadow of doubt on the field for those who remain skeptics.[84]

Advocates and detractors of the theory of endocrine disruption viewed the scientific evidence from starkly different viewpoints. In the most extreme positions, the thesis either sounded the alarm about an important overlooked global health crisis or was a doomsday prediction propagated by environmentalists. Not surprising, *Our Stolen Future* met mixed reviews: it garnered positive responses from the prestigious journal *Science* that, while recognizing most scientists' discomfort with the book's polemical style, agreed that its central premise demanded greater attention from scientists and policy makers and boldly concluded that "the potential threat of endocrine disruptors is a critical issue of our time."[85] The review by Gina Kolata, science writer for the *New York Times,* who received a "sound science in journalism award" from TASSC in 1995, was decidedly negative.[86] The book had been written for a lay audience, and despite the skepticism of the broader scientific community, it met with popular success. It was eventually published in fourteen languages, and its authors traveled around the world to discuss endocrine disruption.

Was endocrine disruption a legitimate concern? What were the central components of the thesis? For the EPA to develop a process to regulate these chemicals, as promulgated by the FQPA and the SDWA, some consensus on the definition of an endocrine disruptor needed to be reached. If endocrine disruption was nothing new to toxicology, few new tests would be required. But if a new thesis of how chemicals interfere with development and manifest in abnormalities and disease as organisms age proved to be legitimate, the testing protocols and the determination of safety, especially at low doses, for many chemicals would require significant revision. Defining endocrine disruption would prove to be a contentious public process as the EPA took up the task of developing a testing and screening program.

DEFINING ENDOCRINE DISRUPTION WITHIN
THE REGULATORY CONTEXT

While it was impossible to deny that some industrial chemicals could, in fact, interact with hormones, the implication of such biological action remained the central issue in the debate over the risks and safety of endocrine disruptors. Were these chemicals a real threat to human and ecological health? Were their risks being overstated to garner public attention and a political response? Was industrial chemical production irrevocably damaging human health for generations to come? Were the risks of endocrine disruptors greater than those of naturally occurring hormones? And if a chemical binds to a hormone receptor, is it necessarily an endocrine disruptor per se, or must such a determination be conditional upon the adversity of effect?

Two high-profile committees, formed to review the issue of endocrine disruption, took up many of these questions: the NRC Committee on Hormonally Active Agents in the Environment, formed immediately before the passage of the FQPA and SDWA (referred to here as the NRC committee), and the EPA's federally chartered Endocrine Disruptor Screening and Testing Advisory Committee (EDSTAC).[87] The formation of these two committees offered representatives from the federal regulatory agencies, private industry, academic researchers, and environmental organizations their first opportunity to hammer out the meaning of endocrine disruption and the extent and scope of the testing and screening required to determine safety. Both processes were fraught with fundamental disagreements about the meaning of endocrine disruption.

The EDSTAC faced a more politically daunting task than the NRC committee, as it was charged with developing recommendations for regulatory testing with direct implications for chemical producers. Moreover, as a federally chartered committee, its stakeholders represented diverse interests, including the EPA and other regulatory agencies, private industry, public interest and environmental groups, and university research scientists. The process was long, laborious, and deeply contentious. In all, the committee held ten public meetings at various locations across the country over the span of close to two years.[88] The EDSTAC had to determine the scope of program (to include evaluation of interaction with estrogen, androgen, and thyroid hormones), a process to screen and test for endocrine disruptors, and a means to prioritize how chemicals would be subjected to testing.[89] The first step in this process required a shared definition of an endocrine disruptor (or *disrupter*, as even the spelling came into dispute). This seemingly simple starting point took the EDSTAC nearly five months to resolve and nearly destroyed the consensus-making process.

Agreeing on the basic problem definition drove a similar wedge into the NRC committee that also manifested in disputes over language use. In the final report, the NRC referred to this debate as a difference in epistemology that led to divergent interpretations of results, scientific uncertainty, and the significance of risk posed by endocrine-disrupting or "hormonally active agents," the newly adopted term of the NRC.[90] This epistemological division separated many of the original Wingspread participants who sat on the EDSTAC and the NRC committee from their industry counterparts—Colborn on the EDSTAC and Howard Bern, Michael Fry, Louis Guillette, Ana Soto, and Fred vom Saal on the NRC committee, versus representatives from Bayer Corporation, Exxon Corporation, Procter & Gamble, Dow Chemical Company, S. C. Johnson & Son, BASF, and FMC Agricultural Products. The sharp division likely contributed to the frustration among the participants.[91] In its simplest form, the fissure formed as participants debated the scope and magnitude of the problem. For industry representatives on both committees, the problem of endocrine disruption was narrow in scope: Do chemicals interacting with the hormone system elicit toxicologically defined adverse effects? In contrast, public health and environmental advocates sought to expand the definition to include considerations for the sensitive timing of exposure and argued that traditional high-dosing toxicity testing had proved to be inadequate in evaluating risk. Endocrine disruption was about more than toxic effects.

Industry representatives stood firm against a broad definition that might suggest that any chemical with endocrinelike effects was an endocrine disruptor and hazard per se. Dose still determined the effect, and safety of endocrine disruptors could, therefore, be conducted through traditional toxicity testing. An endocrine-disrupting chemical must "cause adverse effects," industry representatives concluded. To fail to include such phrasing on adversity, one representative of small businesses argued, would be "intellectually dishonest."[92] In sum, the industry had two objectives: first, to maintain that any interpretation of an endocrine disruptor would be based on a *de minimus* standard and not a per se assumption about endocrine activity; and, second, that the adverse effects evaluated should remain narrowly defined according to traditional toxicity endpoints. Like a strict interpretation of the Delaney clause, a per se definition of an endocrine-disrupting chemical risked overwhelming the regulatory agencies and restricting many economically valuable chemicals without an evaluation of the scope or magnitude of the potential problem. In a letter to the EPA in

1998, the Chemical Manufacturers Association "strongly recommend[ed] that the definition of endocrine disruptor clearly incorporate the traditional toxicological definition of adverse effects to avoid confusing the public about this important scientific concept."[93] The SPI, in its comments to the EPA the same year, echoed this argument: the industry "emphatically maintain[ed]" that the EPA should not classify a chemical with a positive screen as an endocrine disruptor but should uphold the definition that an endocrine disruptor must "cause adverse effects" as shown in traditional toxicity testing.[94]

Implying that endocrine-active agents could be determined via the same study protocols used in traditional regulatory toxicity testing, Wingspread participants argued, entirely failed to account for the central problem of endocrine disruption. Low doses, sometimes below toxic thresholds, in developing organisms could elicit complex effects on multiple systems beyond cancerous tumors and toxicity. Moreover, there might be different adverse effects at low doses than at high doses; therefore, one couldn't simply use high-dose testing to predict low-dose effects and safety. That hormones could elicit strong developmental effects at low doses was perceived as common sense by researchers such as Ana Soto and Fred vom Saal.

Frustrated by what they perceived to be the NRC's inadequate and skewed report, vom Saal and Soto submitted minority reports of dissent. They argued that the NRC report too narrowly defined endocrine disruption by focusing on hormone activity as a functional issue (i.e., does a chemical bind to a hormone receptor or not?) with uncertain adverse effects and ignored entirely the issue of timing of exposure. They contended that by doing so, the committee missed the most troubling aspects of the issue: exposure early in development to even low levels of endocrine disruptors could cause developmental and transgenerational effects.[95]

Mediating between the conflicting positions on the meaning of endocrine disruptors on the EDSTAC was Lynn Goldman, assistant administrator at the EPA's Office of Prevention, Pesticides and Toxic Substances. Goldman, a young, bright, and energetic medical doctor and epidemiologist, was appointed to the position by Clinton in 1993. From the first day of her appointment, endocrine disruption was a high-priority issue for her office. "Such was the sense of urgency attached to the issue," she writes, "that, one day in October 1993, EPA Administrator Carol Browner was asked to swear me into office on an urgent basis so I could testify before a hearing . . . about estrogens in the environment."[96] Goldman, who had studied the effects of pesticide exposure in children, was alarmed by the testimonies of the Wingspread scientists. While she recognized the need for much more research on the subject, she also realized that the scope of the problem extended beyond estrogenic chemicals. With that in mind, Goldman expanded the committee's mandate to include the hormonelike effects of thyroid and androgen-mimicking or -blocking chemicals in addition to estrogenic compounds.[97]

In a memo to EDSTAC members, Goldman sought to hold together a committee that was on the verge of collapse. "The central points of disagreement," she explained, "revolved around whether a definition should be limiting or expansive, and what kind of proof is necessary to label a chemical as an endocrine disruptor." In an effort to strike a balance, Goldman proposed two definitions for an endocrine disruptor: a "potential endocrine disruptor," a category that would include those chemicals that screened positively for hormone activity, and an "endocrine disruptor," a category encompassing chemicals shown to cause adverse effects in more extensive animal testing.[98]

Representatives of industry were livid, and they resolutely rejected this proposal, suggesting that it "risk[ed] destroying consensus on the entire program." The industry's concerns lay with the potential effects of such definitions on future liability. Suggesting that a positive screen for endocrine activity rendered a chemical a "potential endocrine disruptor," industry representatives argued, could propagate "unnecessary public perception concerns, inappropriate product de-selection and unwarranted litigation."[99] They went on to contend that because screening tests were designed to err on the side of false positives rather than false negatives, the screening process could misrepresent some substances as endocrine disruptors when in fact they were not (false positives), with potentially unnecessary economic impacts. (The NRC's final report, *Hormonally Active Agents in the Environment*, contradicted this claim, stating that simple screening methods might result in either false positives *or* false negatives. For this reason, the committee recommended the use of multiple methods of screening.)[100] Goldman's proposed category, from the perspective of industry, created a potential opening for future liability by expanding rather than limiting the definition's scope. Implicit in industry's refusal to accept Goldman's proposal was the recognition that labeling a chemical with an initial positive screen for hormone activity stepped closer to defining a chemical as an endocrine disruptor per se. This necessarily risked eliminating the *de minimus* standard for determining safety and thus the contention that considering dose was absolutely necessary in evaluating risk and safety.

Industry representatives drew their line in the sand. If Goldman wanted to maintain consensus in the committee, she needed to back down from her proposal and maintain the phrase "causes adverse effects" in any definition of endocrine disruptor. Goldman responded to the pressure without entirely conceding to industry's position. Her intention was not to suggest that a chemical was an endocrine disruptor per se if it tested positive in a simple screen using only cell culture (i.e., in vitro). She understood endocrine disruption as a mode of action; yes, a chemical might bind to an estrogen receptor, and yes, this was of concern, but in order to make a regulatory decision on a chemical, adversity needed to be determined.[101]

The final definition agreed to by the EDSTAC illustrated this consensus and the middle-ground position negotiated by Goldman. An endocrine disruptor was described as "an exogenous chemical substance or mixture that alters the structure or function(s) of the endocrine system and causes adverse effects at the level of the organism, its progeny, populations, or subpopulations or organisms based on scientific principles, data, weight-of-evidence, and the precautionary principle."[102] The inclusion of "scientific principles, data, weight-of-evidence, and the precautionary principle" reflected the disagreement over how evidence of endocrine disruption would be used in decision making. Environmental advocates on the committee pushed for the inclusion of the "precautionary principle," while industry insisted on "scientific principles." Writing to the EPA, representatives of the SPI argued that the precautionary principle "is based on certain political and ethical assumptions or beliefs that are independent of any scientific inquiry into facts." The SPI charged that the precautionary principle would interfere with "sound science research," which "SPI and others have made substantial investment in."[103] The argument that sound science and the weight of the evidence had failed to demonstrate that endocrine disruptors presented serious risks to human health would, in the coming years, become an important part of the debate over the safety of chemicals such as BPA.

In the end, however, the EDSTAC definition was remarkably similar to that of the NRC committee. "In its simplest form, the hypothesis is that some chemicals in the environment mimic estrogens (and other sex-hormones) and hence interfere with (disrupt) endogenous endocrine systems, with adverse effects."[104] Subsequent efforts by European authoritative bodies to define endocrine disruption resulted in somewhat similar statements. The World Health Organization's definition in 2002 defined an endocrine disruptor as an "an exogenous substance or mixture that alters function(s) of the endocrine system and consequently causes adverse health effects in an intact organism, or its progeny, or (sub)populations." A slightly different, broader definition was adopted under the Canadian Environmental Protection Act in 1999: an endocrine disruptor is "a substance having the ability to disrupt the synthesis, secretion, transport, binding, action or elimination of natural hormones in an organism, or its progeny, that are responsible for the maintenance of homeostasis, reproduction, development, or behavior of the organism."[105]

The two U.S. committees outlined similar processes to screen and test endocrine disruptors. Chemicals would be prioritized on the basis of exposure and hazard information and then screened for hormonal activity with rapid cell-based tests, referred to as Tier 1 tests. The results of the Tier 1 screening would determine whether a chemical would move on to more extensive and expensive animal studies—or Tier 2 tests—or whether no further testing would be done. The results of the Tier 2 tests would then be used to develop an assessment of a chemical's hazards. The EDSTAC recommended that the Tier 2 tests should consider low doses, which it specified should not be determined simply from high-dose tests.[106]

In the wake of junk science accusations and despite deeply divided participants, both committees concluded that there was enough evidence of concern for endocrine disruptors. At a minimum, everyone could agree that more research was needed. Yet considerable confusion and disagreement remained about the scope of endocrine disruptors, especially the role of low-dose exposures, the importance of the timing of exposure, and, importantly, the meaning of and process to define an adverse effect. Hoping to resolve some of this debate, in 1999 the EPA asked the National Toxicology Program to evaluate the evidence for low-dose effects of estrogens and a number of synthetic estrogens, among them BPA. By the late 1990s, research on this chemical was becoming a lightning rod for the larger political and scientific debate over the safety of endocrine disruptors.

The release of the EDSTAC and NRC reports in 1998 and 1999 put in motion two distinct efforts to evaluate and manage endocrine-disrupting chemicals. The first approach was to develop and validate EPA screening and testing procedures for regulatory purposes. After the EDSTAC released its report, the EPA established a second committee dedicated to this work, the Endocrine Disruptor Methods Validation Subcommittee. This work of validating methods dragged on for over a decade. By the end of George W. Bush's administration, the EPA had not yet tested a single chemical as part of its Endocrine Disruptor Screening Program. Environmental advocates and several of the researchers involved in the EDSTAC meetings largely turned away from the validation process, which they perceived as flawed and painfully slow.

The second response was to test and refine the scientific thesis and build a scientific discipline. As urged by the NRC report, researchers in the United States, Europe, Japan, and increasingly China investigated the biological mechanisms of action (i.e., how a chemical functions at the most basic cellular level to elicit an effect), different dose-response relationships, and the long-term health risks of exposure, particularly at low levels. Scientific meetings continued, including the first Gordon Research Conference on Environmental Endocrine Disruptors in 1998, launching a prestigious international scientific conference series. Researchers focused on multiple chemicals, identified new modes of action expanding well beyond the estrogen receptor, grappled with the problem of evaluating multiple exposures, and began evaluating the interaction of chemicals with the human genome.

This proliferation of research outstripped the development of a well-defined theory of causation. For example, the identification of new modes of action (e.g., binding to thyroid hormones, interfering with hormone synthesis) and new effects measured very far upstream of discrete negative health outcome (e.g., altering insulin regulation, fat cell development, chromosomal alignment) complicated the process of identifying a simple causal relationship between a given chemical and an adverse outcome. In this way, the contest over the thesis of endocrine disruption fit the Kuhnian model of a paradigm shift, wherein scientific thought often emerges even before rules and assumptions are established and, indeed, often guides research long before a reduced set of rules is created.[107]

When the expansion of research crashed up against incomplete and contested theory, conflict erupted, with regulators, industry, and researchers battling over the severity of the threat posed by endocrine disruption. At the heart of this new science policy debate is the question: Could endocrine disruptors be assessed according to traditional toxicology methods and standardized protocols, or did testing need to be significantly reformed? This struggle to define the scope and severity of the problem of endocrine disruption, and therefore an appropriate regulatory response for measuring risk and defining safety, played out in the public debate over the safety of BPA.

5   The Low-Dose Debate

Before the meeting at the Wingspread Conference Center in 1991, chemicals such as BPA had been of no particular research interest to Fred vom Saal. The research that first intrigued him as a young graduate student in neurobiology at Rutgers University in the 1970s involved the study of behavioral differences in genetically identical strains of mice. If not genetics, what accounted for such variability? This was the classic conundrum in evolutionary biology. Is it genes or environment, or both, that determine an individual's development, health, behavior, and personality?

In the early years of his career, vom Saal studied how the hormonal environment during fetal development affects various phenotypic expressions, such as behavior. If changes in the fetal hormonal environment are correlated with phenotypic variability, how stable or resilient is this system to disruption or perturbation? For instance, what are the potential reproductive and behavioral effects of exposure to varying levels of hormones during fetal development?[1] By the early 1990s, vom Saal had shifted the question slightly to consider what happens to reproductive development when synthetic chemicals that act like or interact with hormones are present in the fetal environment in very small concentrations. It was not such a radical conceptual leap for vom Saal to begin introducing endocrine disruptors, or endocrine-modulating or -active compounds, into his experiments. Like hormones, these are chemicals, albeit synthetically produced ones, that may also be present in the fetal environment and that, one could hypothesize, may influence phenotypic expression and functional changes through mechanisms similar to those of naturally occurring hormones.

While perhaps not a dramatic conceptual change, working with an industrial chemical and publishing data on potential effects introduced political and reputational stakes not often experienced in the world of basic research, as vom Saal would quickly come to realize. Questioning the safety of an economically successful substance such as BPA threatened the stability of its market and, not surprisingly, invited dedicated attention and fierce criticism from the chemical industry. Even further destabilizing was the fact that vom Saal and some other researchers working on endocrine disruptors in the late 1990s explicitly challenged not just the safety of BPA but the entire regulatory method for assessing the risks of chemicals. Put simply, these scientists argued that endocrine disruptors defied the logic of traditional toxicological testing and related assumptions about dose-responses and methods for establishing safety standards because they could have marked effects at very low doses. For this reason, the argument continued, the risk-assessment process failed to adequately evaluate the dangers of potentially endocrine-disrupting chemicals. The controversy over BPA would become a scientific and political flashpoint in the broader critique of chemical regulation. This argument resonated among some environmental and consumer advocates, who mobilized the media and politicians in order to build political pressure for regulatory and legislative reforms.

This chapter begins with publication of vom Saal's first BPA study in 1997, which set off a firestorm of scientific debate and quickly grabbed the attention of the chemical industry.[2] Faced with new questions about BPA's safety, the possibility of liability claims, and demands for fundamentally restructuring how chemical safety was assessed and determined, the chemical industry immediately took a defensive position. Working collectively through the major trade associations, the industry stated that BPA was well tested and unequivocally safe at low doses. In response to emerging research on low-dose effects of BPA, industry representatives questioned the reliability, validity, and relevance of the data. In turn, vom Saal, joined by other academic researchers, adamantly defended the low-dose data and the hypotheses that provided biological plausibility to their findings.

As detailed in this chapter and throughout the rest of the book, the back-and-forth struggle to determine whether BPA was safe was very explicitly influenced by the economic stakes. This is not to say that industry's skepticism about endocrine disruptors and the early research on BPA was solely driven by market concerns. What became evident in the 1990s, as more researchers weighed in on the debate, was the considerable confusion involved in interpreting research on low-dose effects. Differences in scientific training, values, and the assumptions of professional communities agitated this confusion and skepticism. The central divisive issue involved determining how much evidence and which evidence was necessary to warrant regulatory action to limit or ban a chemical's access to the market.

Those skeptical that endocrine disruptors presented a significant human health risk maintained the position, favored by industry, that the burden of proof must be causal—that is, there must be incontrovertible human evidence of an adverse effect, solid evidence of human exposure at levels of concern, and supportive biological plausibility. Such evidence should come from regulatory toxicity studies and large epidemiological studies. Conversely, those who viewed the emerging research on endocrine disruptors as indicative of a significant risk or threat to human health advocated lower burdens of proof for limiting public exposure, recognizing the limits and complexity of animal research and epidemiological studies. The issue of causality was the fulcrum. Given the complexity of biological mechanisms and disease development, proving that a chemical directly and unequivocally causes a disease is extremely difficult. The difficulty is only compounded when studies examine increasingly smaller doses and the effects measured occur upstream of the manifestation of disease. Working from the assumption that certainty of causality is rarely confirmed, advocates have called for precautionary approaches to potentially endocrine-disrupting chemicals. Such measures have included limited bans and demands for alternative chemicals or processes. Making public health decisions under uncertainty, however, has proven time and time again to be politically challenging, as demonstrated in the history of DES, lead, and tobacco. Bans on these chemicals and products occurred only after decades of research, political fights, and ultimately damning epidemiological research.

Because skepticism and scientific uncertainty ran high in the early years of endocrine-disruptor research, it was not easy to distinguish when scientific uncertainty emerged from a fair evaluation of research and differences in interpretation (i.e., where the burden of proof lay) and when it was deliberately and knowingly manufactured with the explicit purpose of defending markets and products. Most research on endocrine disruptors and BPA fell into the gray area between sponsored efforts to project shadows of doubt to defend hazardous products and legitimate questions about what constituted valid and reliable research. Faced with both scientific uncertainty and some reason for concern, government researchers at the National Toxicology Program (NTP) called for more research on BPA in 2001. Ten years later, thousands of papers had been published on research conducted in laboratories around the world. The safety of BPA and its markets were destabilized.

VOM SAAL'S RESEARCH SETS OFF THE DEBATE

In the 1980s, vom Saal published several papers on what he called the "womb effect."[3] In them, he described how exposure to different levels of estrogen and testosterone during fetal development resulted in measurable differences in sexual behaviors and reproductive development in mice. It was this research that caught the attention of Theo Colborn when she was conducting research for the *Great Lakes, Great Legacy?* report, and it is described in *Our Stolen Future* as a critical piece of the endocrine-disruptor thesis.[4] In mice, which give birth to litters, pups develop side by side in the mother's womb, forming a horn shape. Whether a given pup's sibling neighbors are male, female, or one of each—sandwiched between a brother and a sister—determines the variability in that pup's exposure to estrogen and testosterone. Positioning in the womb, and thus different hormone exposure there, vom Saal documented, could predict a number of hormonally regulated endpoints, including reproductive behavior (such as aggressiveness), onset of puberty (and length of estrus or menstrual cycle), and body weight.[5] The phenomenon observed by vom Saal suggests that the endocrine system plays a critical role in determining phenotypic variability in a population. This variability increases the "possibility that some offspring in a litter will have a phenotype that is well adapted for the environment in which they live, compete and reproduce."[6]

When he first met Colborn and others at the Wingspread Conference, vom Saal was introducing natural hormones at very minute concentrations during development and observing the effects on a variety of endpoints known to be sensitive to hormonal change. The exposure levels used were in the parts per billion and parts per trillion. These are levels determined to be "physiologically active," or the concentration range at which binding to receptors occurs.[7]

Vom Saal conducted this work in collaboration with Wade Welshons, a colleague in the Department of Veterinary Biomedical Science at the University of Missouri. A Harvard-trained scientist, Welshons had studied biochemistry and oncology as a postdoctoral fellow under Jack Gorski at the University of Wisconsin.[8] Gorski, a giant in the field of steroid hormone biochemistry (including sex hormones), molecular biology, and steroid hormone function, was the first researcher to identify the molecular structure of the estrogen receptor in the 1960s. His work laid the foundation for understanding the mechanisms of estrogen action and associated biochemistry. For example, how do hormones interact within the cell and with DNA? How are they transported in the body to deliver messages?[9] It was Gorski with whom vom Saal wrote the chapter on low-dose effects for the National Research Council (NRC) report on hormonally active agents in the environment published in 1999.[10]

In the 1970s, Gorski's laboratory demonstrated that the relationship among hormone level, receptors, and physiological response operates in a feedback loop system.[11] A very basic understanding of hormones is as follows. Hormones elicit their effects by binding to hormone receptors either on the surface of the cell membrane or in the nucleus. Hormone binding triggers a series of intracellular events that turn genes on and off, producing or suppressing protein production. What emerged from decades of research into understanding this system was the spare receptor theory. The theory argues that the maximal response (e.g., cell proliferation) occurs when hormones occupy—that is, when hormones bind to the receptor protein of—more than 1 percent but less than half of all hormone receptors. The additional receptors above this level are referred to as spare receptors. When hormone levels continue to rise above the point of maximal response, the effect (such as cell proliferation) begins to decrease. This process is referred to as "receptor downregulation."[12] For example, when estrogen is given to a rat whose ovaries have been removed, cells in the uterus proliferate. At some point, when less than half of the receptors are occupied, maximal response is reached. If one continues to add estrogen after the saturation point is reached, proliferation begins to drop off.

The graphical representation of the spare receptor or receptor downregulation is a nonlinear curve.[13] The response increases with increasing dose (positive slope) until the saturation point is reached, after which the response decreases with increasing dose (negative slope). This inverted dose-response curve is referred to as nonmonotonic. With a non-monotonic dose-response curve, the direction of the slope or direction of the curve changes from positive to negative or vice versa. The resulting shape looks like a U or inverted U.[14] The spare receptor hypothesis supports the thesis that the estrogen system evolved to be highly sensitive to very minute changes in hormone concentrations.[15] Consider that the endocrine system functions somewhat like a thermostat that responds to even subtle changes in the environment, turning heat on when the temperature cools below a set point and turning it off when the temperature rises above that set point. The implication for endocrine-active agents is that the dose-response relationship may be non-monotonic, thereby making simple extrapolations about low-dose effects from high-dose effects increasingly complicated.

If hormone signaling evolved to respond to changes in the environment, what happens when that signaling is altered by the addition of novel chemicals, or by endocrine-disrupting chemicals? Is the system resilient, does it change, and, if so, will the changes have adverse impacts on human and wildlife populations? Evolutionary developmental biologists working on the gene structure and binding properties of the estrogen receptor, notably Joe Thornton at the University of Oregon, have hypothesized that the estrogen system among vertebrates has changed very little over the past three hundred million years. Moreover, the estrogen receptor is considered to be somewhat promiscuous, allowing many different chemical structures to bind, block, or partially bind. Such flexibility may be critical to evolutionary adaptation but may also help explain why many human-made chemicals can interact with the receptor.[16] Determining whether this conservancy and promiscuity render humans and other species resilient or vulnerable to the rapid introduction of new sources of estrogens and endocrinelike compounds over the past half century is a tremendously difficult challenge and one fraught with political debate, as the unfolding argument about BPA's risks demonstrates.

After returning from the Wingspread meeting in 1991, vom Saal joined a small number of his new colleagues in exploring the question of resiliency, which lay at the heart of the emergent field of environmental endocrine disruptors. In a series of experiments, vom Saal's laboratory tested the effects of three known estrogenic compounds, DES, BPA, and octylphenol (a compound also used in plastic production), on male reproductive development. The researchers exposed pregnant mice during a critical window in the fetal gestational process (days 11 through 17) to extraordinarily low concentrations of the estrogenic compounds. Pregnant animals were exposed to 0, 0.02, 0.2, and 20 nanograms per kilogram of body weight per day (parts per trillion) DES, and 2 and 20 micrograms per kilogram (μg/kg) of body weight per day (parts per billion) BPA and octylphenol. The levels used for BPA and octylphenol were higher than those used for DES because they were less potent estrogens than DES.[17] In 1997 and 1998, vom Saal's team published several papers from these studies that found significant effects of DES, octylphenol, and BPA on the male reproductive system, including changes in prostate weights and decreased efficiency of sperm production. They reported an "inverted U" or non-monotonic dose-response relationship for prostate weights in exposed animals.[18]

The doses used in vom Saal's study were millions of times lower than any previous work conducted on DES at the time and unprecedented in toxicity studies of BPA. Such low concentrations were selected to test the effects of exposure to physiologically active levels of these estrogenic compounds. In the fetal mouse, the physiologically active level of estrogen is in the parts per trillion (ppt) range—one thousand times smaller than parts per billion (ppb), and a million times smaller than parts per million (ppm). The difference between a toxic and a physiologically active level of estrogen is a hundred millionfold. Vom Saal explained the significance of this dosing difference as it affects chemical research: "So when you have a physiologist thinking of a millionth of a gram, you have that physiologist thinking this is a toxic high dose. When you are raised in the field of toxicology, you are looking at that from the other perspective of 'My gosh, that's such a tiny dose, it couldn't do anything.'"[19]

Compared to reproductive toxicity tests such as the study conducted on BPA in the mid-1980s by James Lamb at the National Institute of Environmental Health Sciences, the doses in vom Saal's research were millions of times lower, and the exposure window was very narrow rather than continuous throughout pregnancy. The objective of this design was to expose the animal precisely during a key window of reproductive development so as to isolate the effect on the prostate. The understanding that chemical exposure during critical periods of fetal development could result in severe malformations or miscarriage had long been recognized within developmental toxicology. The fundamental difference in the emerging research on BPA and other endocrine-active agents, as exemplified by vom Saal's work and earlier work by John McLachlan on DES, was the dosing levels, which fell thousands, if not millions, of times below toxic levels. In this regard, the theory of endocrine disruption integrated research on the fetal origins of disease or what in other corners of the scientific world is referred to as the Barker thesis, named for the British epidemiologist David Barker.

In its simplest form, the fetal origins of disease theory argues that in early fetal development, environmental exposures—including nutrition, tobacco smoke, alcohol, air pollution, drugs such as cocaine and caffeine, and chemicals such as heavy metals and endocrine disruptors—can increase susceptibility to a host of chronic diseases later in life by altering how tissues in the body function. As a fetus develops, entire systems of communication pathways between cells and organs are laid down that direct the growth, development, and function of the individual throughout its life. The fetal origins of disease hypothesis suggests that if this original programming is altered during development a newborn may appear normal but that over time, as the individual develops, functional changes may manifest as increased disease development or increased susceptibility to disease development.[20] Unlike in developmental toxicity, the effects of concern are not necessarily severe defects observable at birth. Instead, effects may be subtle functional or developmental changes that don't manifest as disease until much later in life, long after the exposure has occurred.

Since the 1970s, researchers studying coronary heart disease, diabetes, and insulin resistance have focused on the effects of nutrition, alcohol and other drugs, and stress on susceptibility to chronic disease later in life. This research was informed in part by the work of David Barker, who, in the 1950s, studied children born during the Dutch Famine, which occurred near the end of the Second World War. He found that children malnourished in the womb had a higher risk of developing a number of chronic diseases, including hypertension, insulin resistance, type II diabetes, heart disease, and obesity, later in life. Research on the fetal origins of disease was further expanded upon by the significant epidemiological work in the 1970s and 1980s of Mervyn Susser and Zena Stein at Columbia University, who published extensively on the relationship between nutrition during pregnancy and health outcomes of children.[21]

The thesis that BPA could have significant reproductive effects at extremely low doses presented potentially important implications for public health. Not only were these effects reported at levels below the safety standard, but the findings, according to the researchers, threw into question the long-held assumption that BPA was a weak estrogen. Originally, vom Saal's research team had hypothesized that octylphenol would have a greater effect than BPA on the prostate gland because of its higher estrogenic potency as observed in cell-based tests. They too had initially assumed that BPA was a weak estrogen. Yet the study's findings indicated just the opposite.[22] What accounted for the difference in estrogenic potency between the in vitro and ovariectomized tests and vom Saal's whole-animal study?

Part of the answer, concluded Susan Nagel, then a graduate student in vom Saal's laboratory and the lead author of the 1997 published study on BPA, rested in the inability of BPA to bind with the same proteins that bind free estradiol, the naturally occurring form of estrogen. When a hormone is bonded to protein, it is no longer able to bind to receptors. This inability, the researchers proposed, resulted in a higher than expected level of BPA free and available to actively bind with receptors and to exert an estrogenic effect. This new mechanistic understanding, based on experimental observations, raised the question of whether BPA might elicit stronger estrogenic effects than had been previously assumed.[23]

Figure 5. Frederick vom Saal, Heinz Awards, 2010. (Photo credit: L. G. Patterson.)

Further fueling the University of Missouri researchers' concerns about the public health implications of BPA were two studies from the lab of Nicolas Olea, a Spanish chemist at the University of Grenada. Olea found detectable levels of BPA in food cans lined with epoxy resins as well as in the saliva of patients who had received BPA-based epoxy resin dental sealants.[24] The University of Missouri researchers conjectured that the combination of this new human exposure data and evidence of reproductive effects from exposure to minute levels of BPA in lab animals presented a credible reason to reevaluate the safety of the EPA's reference dose of 50 µg/kg body weight/day, established in the late 1980s.[25]

Vom Saal's findings and the call for a new risk assessment for BPA might have gone largely unnoticed had not Theo Colborn pulled them directly into discussions of the testing protocols for endocrine disruptors during the EPA's Endocrine Disruptor Testing and Screening Advisory Committee (EDSTAC) meetings in the late 1990s. Colborn drew on vom Saal's findings, still unpublished at the time, to highlight the need to incorporate low doses and critical windows of development into regulatory tests of endocrine disruptors. Industry representatives on the EDSTAC pushed back strongly against such a proposal, arguing not only that vom

Saal's work hadn't been replicated but also that the effects it described were not readily correlated with long-term adverse health effects.[26] One member of the EDSTAC summarily dismissed vom Saal's data as a misguided attempt to hold up a flawed theory of endocrine disruption that had been badly tarnished by the embarrassing withdrawal of McLachlan's study on the synergistic effects of estrogens.[27] Viewed in this doubtful light, vom Saal's was just one study, as opposed to part of a larger field of research on environmental estrogens, hormones, and fetal origins of disease.

CONTESTED SCIENCE

The leading BPA producers were quick to replicate vom Saal's research. Even before the publication of his data, vom Saal reported that representatives of Dow Chemical Company had visited him at the University of Missouri to request that he delay publishing his data until the industry had time to replicate his work. Outraged by what he considered to be unscientific and unprofessional behavior, vom Saal wrote up an account of the exchange and sent it to the FDA, Dow, and members of his university. Dow never responded to his letter and subsequently denied the exchange.[28] For vom Saal, the event marked the beginning of his deepening distrust of industry-sponsored research.

In 1997, George Luckett, coordinator of health, safety, and environment at Shell International Chemicals, called for a dramatic increase in funding for research on endocrine disruptors, research that would be done "under the scrutiny of independent scientists," and he highlighted the need to increase funding for communications and advocacy, recognizing that the public would not believe the chemical companies.[29] In response to the BPA findings from vom Saal's lab, chemical producers and trade associations funded a replication study. The Society of the Plastics Industry (SPI) and the European Chemical Industry Council contracted with John Ashby, a researcher at the pharmaceutical company AstraZeneca, to conduct the study. Vom Saal's laboratory team trained Ashby's technicians in the meticulous and difficult procedures involved in dissecting fetal prostates and delivering such tiny doses. The two researchers even spoke frequently by phone over the duration of the study. According to vom Saal, Ashby noted the tremendous difficulty of working at such low doses and with such small organs.[30]

The industry-funded study found no statistically significant effects on the prostate, sperm production or count, or testes in the CD-1 mouse (the same strain that vom Saal's group used) or in rats.[31] Hugh Patrick Toner, the SPI's vice president of technical affairs, stated that the industry's findings put to rest the question of BPA's reproductive effects and, more broadly, "cast further doubt on the broader 'low dose theory.'"[32] A second study sponsored by the SPI and conducted by researchers of major BPA producers, including Shell Chemical, Dow Chemical, General Electric, and Bayer, which was also designed to reproduce vom Saal's research, reported no significant effects on the male reproductive system.[33] By framing the low-dose issue as a simple matter of the veracity of a single study, industry trade groups sidestepped the larger theoretical questions embedded in vom Saal's and other endocrine-disruptor researchers' work, including the fetal origins of disease and the spare receptor thesis used to explain non-monotonic dose-responses. Interpretations of low-dose effects differed significantly between those who viewed the issue through a regulatory lens (e.g., regulatory toxicologists and industry representatives) and researchers such as vom Saal who were trained in endocrine physiology.

A corresponding set of essays, written by James Lamb and vom Saal with his colleague Daniel Sheehan (a biologist from the National Center for Toxicological Research of the FDA), exemplifies the profound interpretative dissonance surrounding the low-dose issue in the mid-1990s. Each essay considered whether risk assessment could adequately evaluate the adverse effects of endocrine disruptors. At the time, Lamb and vom Saal were both serving on the National Research Council's Committee on Hormonally Active Agents in the Environment. Lamb represented the private consulting firm Jellinek, Schwartz, & Connolly, Inc., and in the mid-1980s oversaw the NTP's reproductive toxicity test on BPA.[34]

In his essay, Lamb argued that non-monotonic responses presented nothing new to risk assessment or toxicology. As an example, he drew on a traditional toxicological endpoint: body weight loss. He explained that when exposed to toxic substances some animals immediately died off, which initially allowed the survivors to gain weight. This was then followed by a decline in body weight by the survivors, who eventually died from toxicity.[35] The low-dose issue, from Lamb's perspective, was not about the hypothesis of the spare receptor and downregulation of the hormone system, critical windows of development, or even the determination of the adversity of biological effects at low doses; it was very narrowly and abstractly framed as the existence per se of non-monotonic dose-response relationships in toxicology.

Vom Saal and Sheehan approached the topic from an entirely different perspective, arguing that risk assessment did not adequately evaluate the risks of endocrine disruptors because the method assumed either a threshold or a linear dose-response curve, whereas endocrine disruptors and hormones could elicit non-monotonic dose-responses. They went on to present basic theoretical understandings in endocrinology that supported non-monotonic dose-response curves and explained why significant effects, therefore, might be observed at low doses but not at higher ones.[36] In other words, though they didn't state it this way at the time, endocrine-disrupting effects were not toxic effects. Just as Wilhelm Hueper explained the difference between carcinogenicity and toxicity to members of Congress in the 1950s, vom Saal and his colleague were attempting to make a similar distinction regarding endocrine-active agents. The risks of chemical exposures should not be defined by toxicity alone.

The two essays failed to even frame the problem similarly. They talked past each other, drawing on profoundly different foundations of knowledge to explain non-monotonic dose-response relationships. Lamb viewed the problem strictly through a regulatory lens, conceptualizing the question in terms of whether non-monotonic dose-response relationships per se presented a novel problem to quantitative risk assessment. In doing so, Lamb's logic abstracted the issue entirely from the theoretical discussion outlined by vom Saal and Sheehan and consequently completely ignored their argument regarding the adequacy of current testing methods to evaluate the risks of endocrine-disrupting chemicals. This debate exemplifies the profound conceptual gap between researchers seeking to articulate the risks of endocrine disruptors and regulatory toxicologists in the mid-1990s. They were not even arguing over the same points.

Scientific controversies can have any number of outcomes: they can delay regulatory action, instigate more research, provide fodder for more controversy, or result in all of the above. This cycle of scientific debate, wherein conflict fuels more research and more research fuels conflict, can persist interminably. Industry researchers' failure to replicate vom Saal's findings, combined with the ideological clash over the meaning of low-dose effects, triggered such a cycle of debate and fueled a widening crisis in the consensus on BPA's safety.

DEFINING "LOW-DOSE"

The lack of replication of vom Saal's work by industry studies and the very small body of research on BPA's low-dose effects helped the FDA easily reject calls for a review of the chemical's safety. "Until you can replicate something, you can't interpret its significance," the director of the FDA's

Division of Product Policy, George Pauli, stated in 1999. "Our conclusion is we should go with the track record. We have evaluated [food contact uses of BPA] in a thorough manner, and concluded its use is safe. We haven't seen anything that would persuade us to change that."[37]

While the FDA upheld the industry's position that there was no significant evidence indicating a need to reevaluate BPA's safety, the EPA, under its mandate to develop a testing and screening program for endocrine disruptors, kept the low-dose issue alive. In 2000, faced with profound uncertainty as to how to incorporate low doses into endocrine-disruption testing, the EPA requested that the NTP review the matter. In early 2000, the NTP announced its plan for a three-day workshop, organized by Ronald Melnick, director of Special Programs, to evaluate low-dose research on estrogenic compounds including BPA, DES, ethinylestradiol (estrogen used in birth control), nonylphenol (an industrial compound used in pesticides and plastics and found in drinking water), octylphenol, genistein (a derivative of soy), methoxychlor (a pesticide), estradiol (a natural estrogen), and vinclozolin (a pesticide).[38]

Having spent twenty years at the NTP, Melnick maintained an astute sense of how power, personality, bias, and influence could inform the outcomes of expert panels, and how the initial framing of a scientific question structured process and informed the results.[39] While aware of how the scientific process could be politically manipulated, he held firmly to a commitment to rigorous scientific research and the need to manage distortion, bias, and conflict.[40] Melnick effectively applied this insight to the design of the low-dose review process, which he recognized was already a highly contentious topic. Working with an organizing committee composed predominantly of scientists from the NTP and the EPA, as well as Lynn Goldman, who chaired the EDSTAC process while at the EPA, Melnick created a process that sought to maximize expert knowledge and minimize personal or political influence. For instance, the organizers invited key researchers whose data on BPA were at the center of the scientific controversy to present their research but not to sit on the workshop's decision-making panel. These included vom Saal; Ashby from AstraZeneca; Retha Newbold, who had twenty-five years of experience working with DES at the National Institute of Environmental Health Sciences; Rochelle Tyl, a researcher from Research Triangle Institute, a private laboratory, who conducted several large-scale studies on BPA under industry contracts; and several industry scientists, including Robert Chapin from DuPont Haskell Lab, John Waechter of Dow Chemical Company (a coauthor of several of the plastics industry–sponsored BPA studies), and Frank

Welsh from the Chemical Industry Institute of Toxicology. The BPA panel members responsible for evaluating and interpreting the implications of the research included university researchers and government officials from the United States, Japan, and Canada.[41] The logic of this organizational approach allowed for critical input by the specific investigators but kept these researchers off any of the decision-making panels.[42]

The NTP panel explicitly decided to define the phrase *low dose* very broadly, recognizing that there was no generally agreed-upon definition. Low-dose effects were defined as "biologic changes that occur in the range of human exposures or at doses lower than those used in the standard testing paradigm of the U.S. EPA for evaluating reproductive and developmental toxicity."[43] The choice of the term *biologic changes* implied that a low-dose effect might not necessarily be adverse. They based this decision on the understanding that in many cases the long-term manifestations of physiological changes (e.g., hormone levels or gene expression) reported in several of the low-dose studies had yet to be investigated, rendering any evaluation of the adversity of effects impossible. In other words, lack of research on discrete adverse effects from low-dose exposures would not be conflated with evidence of no effects. In adopting this definition, the NTP made explicit that "effects" included a much broader set of endpoints than those used in standard toxicity tests.

In its final report, the NTP concluded: "Low dose effects were clearly demonstrated for estradiol and several other estrogenic compounds. The shape of the dose-response curves for effects of estrogenic compounds varies with the end point and the dosing regimen."[44] The panel pointed to the low-dose effects of methoxychlor, nonylphenol, and genistein. Traditional toxicity testing methods, they went on to state, inadequately evaluate estrogenic effects because the dosing levels used were within a toxic range. Moreover, predicting low-dose effects from high-dose testing was not appropriate for estrogenic compounds. Quite boldly, the NTP panel called for a reassessment of "the current testing paradigm used for assessments of reproductive and developmental toxicity . . . to see if changes are needed regarding dose selection, animal model selection, age when animals are evaluated, and the end points being measured following exposure to endocrine-active agents."[45] For the first time, the NTP's panel of experts lent scientific legitimacy to the study of low-dose effects and, importantly, addressed the possible inadequacy of current testing models to evaluate the risks of environmental estrogens.

As for the specific review of BPA, the panel came to an ambiguous conclusion reflective of the very limited research available at the time. They stated that "several studies provide credible evidence for low dose effects of bisphenol A."[46] These included laboratory reports of increased prostate weight and advanced puberty, as well as effects on uterine growth and serum prolactin levels. At the same time, they found Ashby's study, which failed to replicate these findings, credible as well. This inconsistency in reported effects, the panel concluded, might have resulted from a number of factors, including the use of distinct species and strains of animals, study conditions, and the variability in the diet of the animals. Estrogen sensitivity varies across species and strain and can be tissue dependent within each species and strain. As for diet, different commercial sources of feed contain variable levels of soy, which is also known to elicit estrogenic properties. Then there is the matter of caging materials, including the use of polycarbonate cages and water bottles, as well as consideration of the timing of exposure as it relates to the endpoint measured.[47] Further, given the subtle endpoints being measured—imagine how tiny a fetal mouse prostate is—and indications that BPA might not simply function like estrogen, replication proved challenging. All of these noted considerations made evident the tremendous challenge in studying the effects of chemicals at such low concentrations.

The significant complexity and specificity in the low-dose research provided abundant opportunity for interpretation and for generating confusion and doubt. The NTP report declared that there was ample evidence of low-dose effects of several environmental estrogens, but it didn't declare that these were adverse effects. On the BPA research, the findings were entirely inconclusive except on one point: the "current testing paradigm" needed reconsideration. Recognizing that the EPA had asked for the report and that the findings could influence the regulatory agency's testing and screening program for endocrine disruptors, industry trade associations and a handful of academic researchers and environmental advocates sought to highlight the flaws and successes of the report.

The major trade associations were deeply unsettled by the report. In public comments to the NTP, the American Plastics Council (APC), the American Chemistry Council (formerly the Chemical Manufacturers Association), and the American Crop Protection Association strongly criticized the NTP's process. They expressed their concern that the EPA might interpret these findings as supportive of low-dose testing for endocrine disruptors and thus as a recommendation to expand or reform testing requirements.[48] And indeed, the NTP had suggested some reforms of the testing protocols. The EDSTAC estimated that endocrine-disruptor testing for a single compound would cost anywhere from $200 to $30,000.[49]

Industry representatives, as demonstrated during the EDSTAC debates discussed in the previous chapter, were intent on narrowing the scope of the testing protocol and the definition of *endocrine disruptor* in order to control projected testing costs and to limit any negative impacts on a chemical's market. The NTP's expansion of the definition of *low dose* to include biological effects or changes, rather than simply adverse outcomes, threatened to expand the testing parameters and costs. For these trade associations, the appropriate scope of the issue before the NTP was whether these compounds elicited toxic effects at low doses and whether there was overwhelming evidence of non-monotonic dose-response relationships. The answer was, of course, "no," because the issue was about, not the toxicity of low doses, but their biological effects on the endocrine system and whether such changes manifested as disease or dysfunction. Such cognitive dissonance between the trade associations' framing of low-dose effects and the NTP's was strikingly similar to the debate between Lamb and vom Saal and his colleague Sheehan.

Colborn and vom Saal completely contradicted the industry's critiques of the report. They interpreted the findings as strong validation of their calls for low-dose testing of endocrine disruptors by the EPA and the inadequacy of traditional toxicity testing.[50] The NTP's conclusions provided support for their conviction that there was an "ongoing 'paradigm shift' that has been occurring in toxicology due to the discovery of endocrine disruption as a mechanism of toxicity."[51] This new paradigm, they contended, demanded that the EPA integrate low-dose developmental testing into its screening and testing program for endocrine disruptors.

The implications of the report reflected both of the critiques. The NTP had called for more research and for reassessment of testing protocols for low-dose effects of endocrine-active agents. It did not outline what those tests should be, as this was not the intent of the report. Nor did the report define which low-dose effects would be determined to be adverse health risks. The implication was that ambiguity remained in the determination of how to evaluate the risks of low-dose exposures to environmental estrogens and which effects should be used to evaluate those risks. Both of these decisions would need to be made for the EPA to integrate low-dose testing into its Endocrine Disruptor Screening Program.

The NTP released its report in 2001, and the following year, the new EPA administrator, Christine Todd Whitman, appointed by George W. Bush, announced that the EPA would not include low-dose considerations in testing and screening protocols for endocrine disruptors. Because of remaining scientific uncertainty about low-dose effects, the agency determined that it would be "premature to require routine testing of substances for low dose effects in the Endocrine Disruptor Screening Program."[52] The EPA's announcement signaled the agency's considerable lack of investment or interest in moving the endocrine-disruptor testing process forward. Indeed, validation of testing and screening protocols and testing criteria (e.g., moving from a screen to more extensive testing) slowed to a snail's pace. Public interest groups that had been actively engaged around the EDSTAC process became increasingly preoccupied by efforts to block potential rollbacks in environmental protection and to disclose alleged scientific manipulation and misconduct within the agencies.[53] Yet while the regulatory process stalled, funding for endocrine-disruptor research expanded, further fueling a widening scientific and political debate over the safety of BPA.

ASSESSING RISK AND THE "WEIGHT OF THE EVIDENCE" ON BPA

The NTP's call for continued research on BPA and its confirmation of some of the low-dose findings were unsettling for chemical producers. Steven Hentges, director of the Polycarbonate Business Unit of the APC, dismissed the NTP's process for evaluating the BPA research as a failure because they did not "complete a weight-of-evidence assessment." Had they done so, he contended, they "would have concluded that low dose effects from BPA have not been demonstrated."[54] What Hentges meant by "weight-of-evidence" turned out to be a very specific evaluative framework and methodology developed at a workshop of the Annapolis Center for Science-Based Public Policy.

Founded by Richard Seibert, formerly vice president of the National Association of Manufacturers, and funded in part by ExxonMobil and Phillip Morris, the Annapolis Center, a nonprofit organization, was formed by "scientists, former policy-makers, and economists who were frustrated by the decision-making process in the environment, health and safety arena."[55] For the center's participants, environmental decision making revolved around what former EPA administrator William Reilly called "episodic panic." Agencies needed to distinguish major risks from minor risks but lacked the tools to do so. The center's mission was to develop those tools for interpreting science for decision makers on hot-button issues.[56] Such issues, as outlined in a strategic planning process in 1998, included clean air, climate change, endangered species, children's health, drug approval ("Is the FDA overly cautious in its approval of drugs?"), and toxicology. In a 1998 letter to Phillip Morris, in which he thanked the company for its $25,000 contribution, Richard Rue, the center's senior vice president, discussed some of the center's major projects. These included a series of workshops and scientific panels to "establish accords for the evaluation of epidemiological and toxicological studies prior to their use to establish regulatory actions by the federal and state governments and our new RegX-pert system (an expert computer system that will allow evaluation of federal and state regulatory actions)."[57]

As outlined in an internal strategy paper, the RegX-Pert system, adopted from a system required by the Department of Defense for scrutinizing weapons systems, was designed to use a series of questions applied in a repetitive way to determine the completeness of a scientific and economic analysis to be used in developing a regulation. Informing the process was the question of whether a regulation was compliant with the cost-benefit requirements of the Office of Management and Budget (OMB). The RegX-Pert system was reportedly developed with input from staff of the House and Senate leadership, staff of relevant congressional committees, and former OMB staff (none of whose names were disclosed in Annapolis Center documents). The program was then reportedly reviewed by George Gray from the Harvard Center for Risk Analysis and Robert Hahn from the American Enterprise Institute and was shared with the OMB for input and feedback. The RegX-Pert system would provide the "benefits" of demonstrating "flaws in the Regs [regulations]" to "the media and Members of Congress," and "for attorneys and judges, it will assist in the adjudication process after the reg [regulation] becomes a law."[58] It is unclear from the limited documentation whether the system was designed to evaluate evidence for making informed regulatory decisions, to challenge regulations, or to do both.[59]

The center was also considering development of an expert witness program. It was "working with General Counsels of major organizations to determine if there is a role for the Center in assisting judges in determining which scientific and/or economic studies are valid and which are not."[60] Put another way, the Annapolis Center was in the business of developing tools to assess and interpret science for decision makers in Congress, the courts, the media, regulators, and the regulators of the regulatory agencies—that is, the officials reviewing regulations at the Office of Information and Regulatory Affairs (OIRA), located within the OMB.

In the late 1990s and early 2000s, a series of workshops called the Annapolis Accords was organized around a number of important themes, including risk (coordinated by John Graham from the Harvard Center for

Risk Analysis), epidemiology in decision making, and cost-benefit analysis (coordinated by Robert Hahn from the American Enterprise Institute and Paul Portney from Resources for the Future). After its workshop series, the center turned its focus to specific issues such as mercury, particulate matter, climate change, asthma, and BPA. In the early 2000s, as David Michaels details in his book *Doubt Is Their Product*, the center produced a number of reports for the coal industry and ExxonMobil on the risks of particulate matter, which cast considerable doubts on proposed regulatory standards.[61] Its forum on toxicology and decision making, held in 2001, sought to address public ignorance on toxicity; as a center planning document stated, just "because a statistical association exists between an alleged hazard and an adverse outcome, the hazard does not necessarily pose enough of a threat that drastic action must occur."[62]

The Annapolis Center, together with the Society of Toxicologists and the American College of Clinical Pharmacology, invited twelve toxicologists to participate in its forum on "toxicology in risk assessment and decision-making."[63] The consensus paper that emerged, the "Annapolis Accords on the Use of Toxicology in Risk Assessment and Decision-Making," outlined criteria for conducting a "weight-of-the-evidence" evaluation. The workshop and subsequent Annapolis Accords report on weight-of-the-evidence methodology were coordinated by George Gray from the Harvard Center for Risk Analysis. (Gray also served as a member of the center's board but resigned the position in 2003.)[64] Embedded within the framework of this methodology was the strong assumption that regulatory toxicology provided the source of "reliable, relevant, and objective scientific information" for risk assessment and that the "fundamental tenet of toxicology" was that "the dose makes the poison." This affirmation of the Paracelsus principle as the proper foundation for risk assessment used in regulatory decision making made explicit the participants' dismissal of the controversial thesis that endocrine-disrupting chemicals might pose risks at low doses not observed at higher, toxic doses. Too often, the authors of the Annapolis Center paper contended, minor risks, such as pesticide residues in food, garner greater public attention than major risks, such as poor nutrition.[65] Low-dose effects of endocrine disruptors would fall into the minor-risk category.

Shortly after the NTP issued its low-dose report in 2001, the APC contracted with the Harvard Center for Risk Analysis to conduct a weight-of-the-evidence assessment using the methodology developed at the center's workshop. The objective of the assessment was "to evaluate the hypothesis that BPA may cause effects in humans at low levels of exposure."[66] This objective was very different from that of the NTP review, which was "aimed at evaluating the scientific evidence on reported low dose effects and dose-response relationships for endocrine disrupting chemicals in mammalian species that pertain to assessment of effect on human health."[67] Determining causality in humans—the central question of the Harvard Center review—would require that an extraordinarily high burden of proof of harm be met. To frame the question as such, given the NTP's recognition that there existed a paucity of epidemiological research and studies on the long-term health effects of endocrine disruptors, was an implicit promise to deliver the very answer industry sought. After all, Steven Hentges of the APC had stated with such confidence that a weight-of-the-evidence assessment would conclude BPA was safe even before it was conducted. Not surprisingly, the report by the Harvard Center confirmed that low-dose effects were not causing harm to humans.

The Harvard Center's assessment of BPA implemented the weight-of-the-evidence criteria by posing two questions. First, was there "evidence for low dose reproductive and developmental effects in experimental animals"? This was answered using three weight-of-the-evidence criteria. First, was there "corroboration" or replication of reported effects, with the presumption that "lack of corroboration is grounds to doubt the validity of single experimental results"? Second, were the studies "rigorous," with greater weight "given to better-conducted studies (e.g., those that use "good laboratory practices)"?[68] Good laboratory practices (GLP) were guidelines adopted by the regulatory agencies in the 1970s in the wake of fraudulent practices discovered at the leading private toxicology laboratory.[69] The third criterion was the statistical "power" at which the findings were reported.

The second question considered whether the effects reported in animal studies could be generalized to humans by evaluating the criteria of the "universality" of the findings (Were the findings restricted to a given species or strain, and were the effects produced in "valid test systems"?) and the "proximity" of effect to humans (Were the effects demonstrated in species similar to humans using appropriate dosing methods, i.e., oral exposure?). Finally, the analysis queried whether the evidence was "consistent with a coherent mode of action" and if that mode was applicable to humans. To answer this, the reviewers determined the "cohesion" of the data—Were they consistent and "subject to a single biologically plausible explanation"?—and the relevance to humans (e.g., similar metabolism).[70]

From the outset, the selection of the questions and the evaluative criteria standards forecasted two conclusions. First, the criteria made explicit that regulatory toxicity testing (i.e., valid test systems and GLP standards) conducted by industry for safety assessments would be more heavily weighted. They would be considered more rigorous and more powerful because of the large numbers of animals used. In practice, this meant that the findings of two large multigenerational studies conducted by Rochelle Tyl at Research Triangle Park on the reproductive toxicity of BPA, which had found no significant effects at low doses, would exert significant influence on the conclusion.[71] The second foregone conclusion was that the analysis would find most of the data reporting low-dose effects (i.e., biological effects) lacking and the effects inconsistent. Low-dose effects had been observed "in much smaller studies not conducted under GLP (good laboratory practices) guidelines," and thus, the report concluded, these findings lacked rigor and statistical power.[72] For those studies that reported effects on a single endpoint—for example, the epididymis (part of the male reproductive system)—the report concluded that "the results do not form a convincing pattern" because some reported increases in weight, others decreases, and some no effect.[73]

As for biological plausibility, "the extent that effects attributable to an agent can all be explained by a single, biologically plausible explanation," the assessment found little convincing evidence. "Differences in the pattern of BPA responses compared to estradiol or diethylstilbestrol (DES)" were found to "cast doubt on estrogenicity as the low dose mechanism of action for BPA," effectively providing poor cohesion.[74] To come to such a conclusion required that one assume that BPA's only mechanism of action was through the estrogen receptor and that this understanding was certain. Yet the report noted multiple mechanisms of action, noting that "data indicate the potential for BPA to have differential in vivo (anti) estrogenic activity that depends on the dose and the tissue undergoing the response." This nuance about mechanistic complexity and the relationship between the dose and the targeted tissue, however, was lost because of the a priori assumption of a "single biologically plausible" explanation—in this case, that BPA acted like a classic estrogen.[75] As Sir Bradford Hill, in his famous 1965 essay on causation, cautioned, biological plausibility is "a feature I am convinced we cannot demand. What is biologically plausible depends upon the biological knowledge of the day."[76] The mechanisms of action of BPA are complex and uncertain, as both the NTP and Harvard Center reports concluded. But in the weight-of-the-evidence methodology, such complexity was interpreted as undermining the cohesion of the data rather than informing the differing observations.

In an odd twist that aimed to further challenge the relevance of low-dose effects on humans, the Harvard Center report pointed to "indirect evidence" that humans were less sensitive to BPA exposure.[77] This latter conclusion drew upon a single paper written by Raphael Witorsch, a professor of physiology at Virginia Commonwealth University School of Medicine.[78] Witorsch argued that because humans produced high circulating levels of estrogen during human pregnancy, exposure to low levels of environmental estrogens, such as BPA, was unlikely to have adverse effects. The certainty of this statement and the lack of actual data on human exposure to BPA during pregnancy were inconsistent with the high burden of proof for causality applied to the rest of the data. Indeed, fetal exposure to naturally occurring estrogens has not yet been completely demystified.[79] The argument appears to have been drawn for a specific purpose: to throw further doubt on the likelihood that humans might be at risk from BPA exposure.

Witorsch's argument about high levels of circulating estrogen reflected his deep skepticism about the adverse effects of environmental estrogens at low doses and his related affirmation of the resiliency of human development.[80] This skepticism positioned him well to serve in defense of the safety of environmental estrogens, and he made himself available as an expert witness to discuss the "validity" of the relationship between xenoestrogens and adverse health effects.[81] Witorsch had also previously provided expert witness testimony to defend the safety of secondhand smoke for the Tobacco Institute, the tobacco trade association that used scientific manipulation, the manufacturing of doubt, and scientific uncertainty to deny the health risks of cigarette smoke.[82] He served on the board of the Environmental Health Research Foundation along with David Kent of Keller and Heckman LLP, the law firm of Jerome Heckman, which represented the SPI. In a 2003 letter to the editor of *Environmental Health Perspectives*, the Environmental Health Foundation's executive director, John Heinze, resolutely defended the safety of BPA and declared that he had no conflict of interest. Yet he had also been a consultant for the SPI, and his organization received funding from the plastics and chemical industry.[83] Today Heinze works for the public and media relations firm John Adams Associates, Inc., along with Hugh Patrick Toner, formerly with the SPI.[84]

The Harvard Center's assessment concluded that the existing research did not provide reliable and relevant evidence that low-dose exposure to BPA caused harm to humans. Was this assessment an effective product protection or defense method, or did it reflect an independent assessment of the existing data? Distinguishing between legitimate scientific uncertainty and the manufacture of doubt is not an easy task. That there existed considerable uncertainty about whether low doses of BPA caused adverse effects in humans was unequivocal, as discussed in both the NTP's review and the Harvard Center's assessment. But the APC also had many reasons to contract with the Harvard Center, based on its relationships with the Annapolis Center and the Harvard Center's founding director, John Graham (a leading researcher in risk science), and the shared viewpoint these centers and their leaders held on low-dose risks of endocrine disruptors in general. Whether the Harvard Center was a pawn in a deliberate effort to distort scientific interpretation or whether its report was reflective of a distinct ideological approach to evaluating risks independent of industry bias is not simply discernible. Yet it is unequivocal that Graham viewed endocrine disruptors as presenting very marginal public health risks and believed that these risks were easily distorted by public misperception.

THE SCIENCE AND POLITICS OF RISK

In its most basic formulation, risk science is organized around the concept that there are more risks in life than society has the necessary resources to manage and regulate. Given that individuals prioritize risk, not according to a risk's severity or probabilistic outcome, but according to other factors related to risk perception (e.g., familiarity of risk, controllability, time frame), social science methodologies in risk analysis, decision analysis, and risk-risk trade-offs (a comparative model of risks) seek to provide empirical and analytical evidence to guide priorities of the regulatory state.[85] The foundation of the field rests upon the work of leading thinkers in economics in the post–World War II period, such as Kenneth Arrow and James Buchanan, who used economic models to explain political and social choices and developed theories of public choice and rational choice. The central concept of both theories contends that humans' political and economic behavior is rational, with individuals seeking to maximize their material income.[86]

Subsequent research in risk perception has called into question decision making based on rational choice. Individuals appear to make decisions based not simply on actuarial tables and probabilities but also on emotional, cultural, and social biases.[87] Does this mean that the lenses through which the public perceives risk are wrong per se (i.e., individuals and regulators do a poor job of prioritizing risk) and therefore must be corrected through risk analysis? Or must a pluralistic society respect these ingrained tendencies? The question comes down to determining how risks are identified and prioritized (e.g., high probability, low impact; low probability, severe impact). David Gee, formerly of the European Environmental Agency, argues that even those who uphold the strongest version of rational choice theory acknowledge the challenge of comparing distinct preferences in risk acceptance and avoidance.[88]

John Graham studied and worked with leading thinkers in risk perception, risk analysis, and decision analysis throughout his education. At Duke University, where he completed his master's, Graham studied under James Vaupel, a former student of Howard Raiffa. A skilled mathematician, Raiffa made significant contributions to the fields of decision theory and game theory and later taught at the Harvard Business School and the Kennedy School for Public Policy. In the mid-1970s, Raiffa developed an interest in risk-benefit analysis as it related to medicine, nuclear power, and global climate change. In the early 1980s, he advised EPA administrator William Ruckelshaus on the use of risk analysis and decision analysis at the agency. Vaupel, who studied why individuals tend to neglect familiar risks and focus on unfamiliar ones, introduced Graham to Raiffa when Graham first arrived as a postdoctoral fellow at the Harvard School of Public Health in 1984, having just completed his PhD in urban and public affairs at Carnegie-Mellon University.[89] He quickly rose to become deputy chairman of the Department of Health Policy and Management at Harvard in the late 1980s and established the Center for Risk Analysis with funding support from private industry and foundations as well as minimal support from Harvard University.[90]

Graham's support of regulatory reform and utilitarian public health interventions such as a ban on trans fats and seat belt laws put him at odds with libertarian-based organizations such as the Cato Institute. At the same time, his outspoken disdain for the environmental movement and the precautionary principle, along with his embrace of a risk-risk analytical model, put him in an unfavorable light with some environmental organizations, which perceived these methods as underestimating environmental hazards. He found respect and financial support among politically conservative think tanks, business organizations, and industry trade associations. He sat on the boards of Steve Milloy's Advancement of Sound Science Coalition and the American Enterprise Institute–Brookings Institute Joint Center for Regulatory Affairs.[91] His center at Harvard attracted private donations from a long list of companies and trade associations, including the American Chemistry Council, the SPI, Dow Chemical Company, the

Business Roundtable, Phillip Morris, Exxon (the company had not yet merged with Mobil), and General Electric.[92] Members of Graham's executive council in the early 2000s included lawyers from major firms such as King and Spalding, which has represented Dow Chemical, Shell Oil, ExxonMobil, and General Electric, as well as Earnest Deavenport, chairman and CEO of Eastman Chemical, and Jerry Jasinowski, president of the National Association of Manufacturers.[93]

Beginning in the late 1980s, Graham's work joined the growing multidisciplinary field of risk science. The "pioneers in the emerging field," according to Graham and his colleague Jonathan Weiner, include political scientist and legal scholar Cass Sunstein (who later became administrator of the OIRA at the OMB under President Obama); Peter Huber, who coined the phrase *junk science* and argued that the tort system profoundly distorted the penalization of risks; Aaron Wildavsky, the political scientist who wrote a seminal text on the cultural perceptions of risk with anthropologist Mary Douglas; Bruce Ames, the scientist who developed an index for comparing all carcinogen risks, natural and synthetic; and Terry Davies, one of the architects of quantitative risk assessment at the Washington, D.C., think tank Resources for the Future.[94] Many of these risk-science scholars share a similar perception that emotionally driven, irrational decision making by individuals can distort the efficient allocation of limited resources for regulation. As an example, Graham and Sunstein have both argued that emotional responses to a perceived risk adopted within communities of shared values can create a snowball effect, generating a cascade of responses that includes ineffective or inefficient regulatory decision making, particularly when resources are devoted to regulating small, unfamiliar risks rather than large risks. Both point to the banning of DDT as an example of an irrational response (or snowball effect), in this case to Rachel Carson's *Silent Spring*. Sunstein, while acknowledging that Carson contributed greatly to society by raising public awareness of the reproductive toxicity of DDT and the dangers of pesticides, contends that the subsequent response to the risk—the ban on DDT—resulted in other, unaccounted-for risks (e.g., the use of more toxic pesticides) and increased costs. The state responded to emotionally charged arguments by environmentalists when it should have deliberately and rationally accounted for the costs and the magnitude (i.e., quantification) of risk.[95] To avoid emotional cascades, which turn small risks into big ones, Sunstein and Graham present tools, such as cost-benefit analysis and risk-risk analysis, to analytically assess risk in a seemingly more objective manner. Emotional cascades and snowball effects in communities of shared values are not, however, equally applied to chemical or industrial producers, and the influence of power and financial incentives is often absent from these analyses as potential sources of biases in decision making.

Debates about the effectiveness, bias, and equity of these risk tools exemplify what German sociologist Ulrich Beck described in 1992 as the prevailing social and political questions of the "risk society." Whereas in industrial society social and political questions focus on the distribution of productive wealth, in those societies where material needs have been reduced and "exponentially growing productive forces" have introduced "hazards and potential threats," the dominant political and social questions shift to consider the distribution of risk.[96] In this risk society, as detailed throughout this book and confirmed by the risk logic of Graham, chemical exposures have long been posited as marginal and acceptable risks because of the low levels of exposures presumed. From this perspective, concerns about carcinogens in foods, pesticides in the home, or BPA in human bodies are distorted perceptions of risks rather than serious long-term risks to the health of the human population.

Not surprisingly, competing demands for the restrictiveness or leniency of a proposed health or safety regulation, or the distribution of the risk, reflect divergent political ideologies about the appropriate role of the state in market intervention. Scholars on the political left frequently argue that cost-benefit analysis and risk analysis neglect issues of equity in distribution of risk (i.e., who is most vulnerable to exposure?) and fail to acknowledge the paucity of data on costs of health burdens of risks, which results in disproportionately underestimating both costs and risks. The environmental justice movement of the early 1990s drew political attention to profound inequities in risk burden, and today research on disparities in health (including exposure to pollution) continues to raise the importance of the issue of risk inequity.[97] Further, critics of cost-benefit analysis contend that such assessments ignore long-term or future generational impacts of risks in favor of reduced costs in the present.[98] This has serious implications for delayed risks, which may include endocrine disruption or carcinogenesis; both the risks and social burdens of exposure are woefully underestimated.

The precautionary principle, supported by many environmental advocates and an accepted policy provision of the European Union, proposes to provide the means to make decisions that protect public health and the environment in the face of considerable scientific uncertainty. Decisions following the precautionary principle may be based on a lower burden of proof; therefore, in the absence of proof of causality, the state may err on the side of caution when evaluating environmental and health risks.[99] Specifically, the precautionary principle proposes to provide decision-making tools to address problems whose risks have not been defined with discrete, well-defined probabilities. This is not to say that the precautionary principle allows for decision making under ignorance. Rather, the objective is in part to provide guidance in situations where, because of complex systems (e.g., ecosystems, climate, human biology), evidence of causal relationships is exceedingly difficult to obtain, but where there is evidence of irreversible, serious harm to the public's health. In other words, where the risk of drawing a false negative conclusion—for example, that exposure X does not lead to reproductive failure when in fact it does and could result in large-scale population effects—the precautionary principle holds that it is better to proceed with caution than to wait for all the evidence to come in. A number of related assumptions that inform the precautionary principle put it at odds with scholars such as Graham. These include the default assumption that human health and the natural world are vulnerable rather than resilient, and the implicit understanding that profound complexity and variability exist in the biological world that inform a humility in the pursuit of truth through scientific research.[100] More specifically, in taking a precautionary action, one does increase the likelihood that the action will have been unnecessary. (Decreasing the probability of a false negative will increase the probability of a false positive.) However, in taking this risk, the decision maker explicitly puts the public's health or safety above the risk of ineffective allocation of resources.

For Graham, a strong level of causal certainty is required for taking regulatory action. Better to be certain than to take precautions, because acting without absolute certainty can distort the market and the efficient allocation of resources. This approach, however, will inevitably miss early-warning signals. In a speech at the Heritage Foundation, a conservative think tank, in 2003, Graham discussed the "perils of the precautionary principle" as including, at worst, restrictions on innovation and economic progress and, at best, the diversion of regulatory agencies' attention to focus on small, speculative risks. Graham pointed to "exaggerated claims of hazard," such as the disputed link between cell phone use and brain cancer, and "disruption of the endocrine system of the body from multiple, low dose exposures to industrial chemicals," as examples of misguided regulatory priorities. While some might point to tobacco as an exemplary case in which heeding early-warning signals might have saved many lives and resources, Graham urged his audience not to "belittle the scientific complexities" involved in the debate over whether cigarette smoke causes lung cancer, which, he argued, ultimately required large epidemiological studies to resolve.[101]

Graham left the Harvard Center in 2001, when the Bush administration appointed him administrator of the OIRA at the OMB, a position known as the regulatory czar because of its powerful oversight of health and safety regulations. George Gray, who studied as a postdoctoral fellow under Graham, became the acting director of the center and conducted the BPA review. He left the center at the end of 2005, when President Bush appointed him assistant administrator at the EPA's Office of Research and Development.[102]

From the outset, the Annapolis Center weight-of-the-evidence framework that was used to evaluate BPA established causality as the burden of proof necessary for taking regulatory action. But the framework also reproduced prevailing assumptions in risk science: that low doses were marginal public health risks and that endocrine disruption in general presented a distortion of risk. Whether intentional or not, this methodological approach cast doubt on any concerns about the low-dose effects of BPA. Importantly and more subtly, the methodology also cast much of the low-dose research as unreliable and irrelevant to humans—not quite junk science, but not useful in decision making. Conversely, the criteria used upheld standardized regulatory toxicity testing as "sound science" for regulatory decision making.

DEFINING "SOUND SCIENCE" AND THE ART OF NITPICKING EVIDENCE

A second weight-of-the-evidence study of BPA, using the same methodology applied by the Harvard Center, was conducted by the Gradient Corporation several years later, once again under contract with the APC. This second study also came to the same conclusions. The existing research did not present reliable and relevant evidence that BPA exposure at low doses caused adverse effects in humans. Perhaps by happenstance, the Annapolis Accords criteria used in both reports on BPA assessed the reliability and the relevancy of the data, which are the two standards used by a judge to evaluate the legitimacy of evidence as "sound science" before allowing it into the courtroom.[103] With the Daubert decision, judges became the gatekeepers of scientific evidence and, in that position, generated a demand for methods and means to evaluate the weight of the evidence. How else would judges, not often trained in the sciences, assess the evidence? Yet employing legal standards of evidence to evaluate scientific evidence disrupts what Harvard legal scholar and historian of science Sheila Jasanoff calls the "axis of deference" between law and science. "The law in short," Jasanoff contends, "claims to do justice by partially preserving the independent authority of science—by in effect, writing science into the law."[104]

Defining what constitutes sound science includes two practices: first, developing methods to identify relevant and reliable research to inform judges and regulatory decision makers, and, second, identifying junk science or nitpicking evidence. The conclusions drawn from the weight-of-the-evidence assessments of BPA contributed largely to the first objective but also served the second by effectively implying that the majority of the low-dose studies were useless. Legal scholars Thomas McGarity and Wendy Wagner have noted the relationship among the *Daubert* decision, its call for sound science, and the practice of nitpicking research to seed doubt. While *Daubert* might reduce the "number of specious claims," they argue, "it also generated a powerful incentive . . . to manufacture uncertainty."[105]

The second part of defining sound science reflects a more questionable approach to evaluating evidence, one that often involves lawyers and public relations firms rather than scientific experts. The goal is often to protect the product and defend against regulatory or liability actions. The art of picking apart individual scientific studies is a ready means to undermine the reliability or relevance of damaging data. Combined with efforts to define more favorable data as sound science, nitpicking or manufacturing doubt can help protect a product from regulatory oversight.

Understanding this legal strategy and the art of the nitpick requires a short discussion of a man named James Tozzi. Considered the "master craftsman" of antiregulatory forces inside Washington, D.C., Tozzi is (in)famous for his ability to work behind the scenes and quietly kill a proposed regulation without leaving a trace, earning him the nickname "Stealth."[106] In a highly colorful interview with Chris Mooney for his book *The Republican War on Science*, which apparently involved some drinking and jazz piano playing (Tozzi's pastimes), Tozzi summarized his career objective as an effort to "regulate the regulators."[107] Most notable among his antiregulatory achievements are two amendments slipped into large appropriations bills that provided new legal means to nitpick evidence: the Shelby Amendment and the Information Quality Act.

Tozzi first came to Washington, D.C., in the mid-1960s to review regulations at the Army Corps of Engineers. When Richard Nixon came into office, he asked Tozzi to run the "Quality of Life Review," a process outlined in a two-page 1971 presidential memo that directed all health and safety regulations to be evaluated by the White House. With one hand Nixon signed into law legislation that created the EPA and OSHA and dramatically expanded federal oversight of the environment and workplace, and with the other he placed a check on this new authority by giving the newly reorganized Bureau of the Budget, renamed the Office of Management and Budget, centralized oversight of all regulations. Tozzi was there at the inception of this new authority at the OMB and spent the next several decades working inside and outside government to formalize the oversight process into law and practice.[108]

A critical step in formalizing the executive office's authority to review regulations came with the establishment of the OIRA within the OMB. Just before leaving office, President Carter had signed the Paperwork Reduction Act, which, as its title suggests, sought to reduce the burden of government paperwork on the American public, including businesses. The law centralized oversight of information coordination, management, quality, dissemination, and policies for reduction within the OMB and established the OIRA as a clearinghouse for managing and reviewing information requests, including proposed regulations.[109]

When Reagan took office in 1981, he took steps to consolidate oversight of regulations within the OIRA. One of his very first acts as president was to sign Executive Order 12291, which directed all agencies to conduct cost-benefit analysis or Regulatory Impact Analysis for all regulatory proposals. Under the order, the new OIRA, led by James Miller and a small team of economists and lawyers, began to expand its responsibilities for the oversight of regulatory proposals, Regulatory Impact Analyses, and final rules. Despite concerns within Congress about the rapidly expanding authority of this small office, Reagan, at the start of his second term, issued another executive order that allowed the OIRA to intervene even earlier in the process of regulation by requiring agencies to clear even proposed plans for regulation through the OMB. Regulatory proposals and final rules dropped 34 percent during Reagan's first term in office. A panel of the National Academy of Public Administration suggested that this drop in regulation might have resulted from the OIRA's tendency to focus on "nit-picking and minutiae."[110] Such a criticism contradicted the argument that cost-benefit analysis and risk analysis rationalized the distribution of risk priorities and regulatory expenditures.

By effectively implementing President Reagan's antiregulatory agenda, the OIRA came to be known as the black hole where regulations went in for review and never again saw the light of day.[111] Administrator Miller described his job as "being the editor of the *Federal Register*."[112] Tozzi had his hand deep in the process. He recalled to Mooney "with a cackle" that environmentalists wondered, "Christ—who's running EPA—Tozzi?"[113] The legality of the OIRA's regulatory review authority, its lack of public disclosure, and the question of whether it usurped the power invested in agencies by the legislative branch were explored in a number of congressional hearings in the 1980s and eventually were taken up by the Supreme Court. In 1990, in *Dole v. United Steelworkers of America*, the court ruled by a 7–2 majority that the Paperwork Reduction Act did not give the OMB the authority "to review and countermand agency regulations." The case involved the OMB's disapproval of OSHA standards for disclosure of information on chemical hazards to employees in some sectors, based on the argument that disclosure was not necessary to protect employees.[114]

Despite the Supreme Court ruling, Tozzi continued to work quietly through back channels to expand the OMB's power over the scientific review process in regulatory decision making. After serving as deputy administrator of the OIRA, Tozzi left the government and set up a consulting firm in the Dupont Circle neighborhood of Washington, D.C., first under the name of Multinational Business Services, which later became the Federal Focus and is now, in its most recent iteration, the Center for Regulatory Effectiveness. Tozzi's contracting work integrated and formalized strategies used to control the assessment and interpretation of scientific research by regulatory agencies and to beat back unwelcomed regulations for industry clients. He worked with major corporations, most notably Phillip Morris, to protect products and inform the regulatory process. In the mid-1990s, Multinational Business Services contracted with Phillip Morris to fight new regulations on secondhand, or environmental, tobacco smoke. Tozzi worked to obtain raw data from a number of epidemiological studies that demonstrated a correlation between secondhand smoke and increased risk of lung cancer. In a letter to the scientist Elizabeth Fontham, whose research reported significant adverse effects of long-term exposure to environmental tobacco smoke, Tozzi outlined the major flaws in her study and requested the original data for further review by his client.[115] Gaining access to damaging epidemiological data was a growing concern not only for the tobacco industry but also for other major industries facing regulations on air pollution and environmental toxins.

Working with Senator Richard Shelby (R-AZ) in the late 1990s, Tozzi drafted legislative language that would provide a legal tool for data disclosure. The Shelby Amendment, or Data Access Act, slipped into a massive appropriations bill and signed into law by President Clinton, directed the OMB to "to require Federal awarding agencies to ensure that all data produced under an award will be made available to the public through the procedures established under the Freedom of Information Act."[116] This meant any researcher receiving a NIH grant could have his or her data taken for scrutiny even after a study had been peer-reviewed. Data disclosure was of particular interest to Shelby. In an effort to block a 1997 regulation for clean air that would have had serious implications for a major power plant in Arizona, Shelby attempted to force researchers at Harvard University to release their data from a major epidemiological study that found a strong relationship between shortened life expectancy and air pollution in six U.S. cities. The Harvard team refused to release their data after the Health Effects Institute, a nonprofit research organization, and the National Research Council had reviewed and confirmed their findings.[117]

While the Shelby Amendment purported to provide greater transparency in federally funded research, it effectively formalized a well-honed strategy for casting doubt on scientific evidence. In a further step toward institutionalizing sound science practice in regulatory decision making, in 2000 Tozzi drafted the Information Quality Act (commonly referred to as the Data Quality Act) with the input and support of the "experienced hand" of Phillip Morris.[118] The Data Quality Act, much like the Shelby Amendment, was slipped into a giant appropriations bill and signed into law by Clinton with little congressional debate. It directed the OMB to "provide policy and procedural guidance to Federal agencies for ensuring and maximizing the quality, objectivity, utility, and integrity of information (including statistical information) disseminated by Federal agencies in fulfillment of the purposes and provisions of chapter 35 of title 44, United States Code, commonly referred to as the Paperwork Reduction Act [signed into law by Carter]."[119] Under the new law, the OMB became an adjudicator of sound science, despite the paucity of scientists within the office. The law, as McGarity and Wagner argue, provided a "potentially powerful and largely unconstrained weapon for challenging the quality of science (and other information that agencies rely on)."[120] Data Quality Act petitions have been submitted by the American Chemistry Council to challenge the Consumer Product Safety Commission's ban on arsenic-treated wood in children's playground equipment and on the endocrine-disrupting chemicals phthalates and atrazine; the latter two were filed by Tozzi.[121] "Any person can file a petition demanding 'correction' of information 'disseminated' by an agency," Tozzi explained, "including scientific studies, if that person believes the information is not reliable, is not objective, lacks utility, or is biased."[122]

Graham's work as head of the OIRA during the Bush administration pleased Tozzi considerably. Graham "came in, and he did an unbelievable job," Tozzi remarked. "Better than I could have done had I been there myself."[123] While at the OIRA, Graham drew on the Data Quality Act to provide legal grounds for demanding that all regulatory agencies standardize scientific review and risk assessment in order to increase the transparency and quality of scientific information.[124] He enforced a strong policy for peer review of government scientific information and attempted to revamp risk-assessment practice.[125] The first draft of the peer review bulletin was widely criticized as an unnecessary, laborious process that would slow down the regulatory process; moreover, it blocked any academic researcher who received federal funding from serving as a reviewer, which would have effectively disqualified nearly all academic researchers, but included no restrictions for financial conflicts of interest. The revised bulletin removed the conditions for federally funded researchers but, as Mooney contends, did not sufficiently address the problem peer review sought to solve.[126] What it did presume was that the regulatory agencies needed greater oversight and "regulation" by the OMB to improve the efficiency and effectiveness of their decisions. An extensive review by a committee of the National Academy of Sciences declared the OMB's Risk Assessment Bulletin to be "fundamentally flawed" and recommended its withdrawal. In sum, the NRC report found that the document pushed standardization well beyond the scientific evidence, which could result in degradation of the risk-assessment process rather than its improvement.[127]

When does the practice of defining and standardizing the practice of sound science become more than an effort to improve the quality of information and develop into a deliberate effort to restrain environmental and health regulations? When does the practice provide opportunities to defend products in ways that distort the translation of information or undermine independent research? In the case of the weight-of-the-evidence assessment of BPA, purportedly designed to be transparent and objective, sound science proved the chemical safe. The value assumptions, however, made in posing the question of causality and in selecting a priori assumptions about biological plausibility and dose-response relationships (i.e., the dose makes the poison), strongly influenced the outcome. This is inevitable in any assessment. The unsettling aspect of the weight-of-the-evidence methodology is the conflation of large, industry-funded studies that use GLP with better—more reliable and relevant—or sound science for use in decision making. Left out of these standardized approaches to defining objective science are the uncertainties and complexities of the empirical research, as well as distinctions between traditional toxicity tests and much of the emerging research on low-dose effects. The NRC criticized Graham's risk-assessment proposal for going beyond the science; a similar critique applies to weight-of-the-evidence methodology. As in most studies of the health risks of chemical exposures, the tremendous complexities involved in determining the strength of the relationship between exposure and disease provide ample room for doubt. Sound science approaches to standardizing methodologies that hold causality as the necessary burden of proof are welcomed by industry as a means to protect and defend the safety of products.

When the practice of sound science becomes part of a deliberate campaign to defend a product, a more nefarious side of manufacturing uncertainty is revealed. In his book *Doubt Is Their Product,* David Michaels provides extensive evidence of coordinated efforts by "product defense" firms to do exactly what their name suggests: defend a product by manufacturing uncertainty and generating doubt. At a meeting on low-dose effects of chemicals in 2007, a leading product defense lawyer, Terry Quill, strongly recommended the use of weight-of-the-evidence criteria, such as those outlined in the Annapolis Accords, to assess sound science.[128] Vice president of the firm Products Defense, Quill previously had served as senior vice president of the Weinberg Group, another leading product defense firm. Michaels notes that Quill "also has roots in the tobacco wars. . . . He served as outside counsel to Phillip Morris in the secondhand-smoke litigation."[129] When asked where he learned about the methodology, the lawyer claimed he had simply found it on the Internet.[130]

Some of the specific methods and strategies of product defense firms were clearly outlined in a letter from the Weinberg Group to DuPont de Nemours & Company in 2003. The Weinberg Group detailed the scope of its work for DuPont to deter the threat of expanded regulation of and litigation related to perfluorooctanoic acid (PFOA), an industrial surfactant used in Teflon, Gore-Tex, and stain-resistant fabrics.[131] "Specifically, during the initial phase of our engagement by a client, we will harness, focus, and involve the scientific and intellectual capital of our company with one goal in mind—creating the outcome our client desires." Compiling and reviewing the scientific data, Weinberg proposed to develop a "multifaceted plan to take control of the ongoing risk assessment by the EPA, looming regulatory challenges, likely litigation, and almost certain medical monitoring hurdles." Folded into this scientific approach was a public relations arm of the defense, which would involve controlling the perception of the chemical's safety. The letter noted that "blue ribbon panels" of "thought leaders" would be organized in the areas of the manufacturing plants and "areas of likely litigation"; the firm would develop "focus groups of mock jurors to determine the best 'themes' for defense verdicts and perspectives on management of company documents and company conduct." Weinberg suggested it would "reshape the debate by identifying the likely known health benefits of PFOA exposure by analyzing existing data and/or constructing a study to establish not only that PFOA is safe over a range of serum concentration levels, but that it offers real health benefits." It would "coordinate the publishing of white papers on PFOA [and] junk science . . . [and] provide the strategy to illustrate how epidemiological association has little or nothing to do with individual causation." Finally, it would "begin to shape the Daubert standards in ways most beneficial to manufacturers."[132] (In 2005, DuPont agreed to pay the EPA a fine of $10.5 million, and $6.25 million for environmental projects, in exchange for an end to the agency's investigation into whether the company had knowingly withheld information on the risks of PFOA.)[133]

This was the case of PFOA. What about BPA? Was the struggle to manage the safety of BPA caught up in similar efforts to defend a product? Were the weight-of-the-evidence reviews—even if not understood as such by the contractors—components of a campaign to protect the chemical? (The evidence to answer that question, if it exists, would likely sit in a public relations or product defense firm.) It was quite clear, however, by the early 2000s that the struggle to determine BPA's safety had become part of a public relations fight. And as the debate became increasingly polarized, discerning legitimate complexity and uncertainty from manufactured doubt became increasingly challenging. The tenor of the debate became highly contentious and at times deeply personal.[134] For his part, vom Saal's profound skepticism of anything that smacked of industry involvement deepened. He stepped further into the role of the advocate to defend his research and the thesis that low-dose effects were real and of legitimate public health concern.

The fact that the weight-of-the-evidence reviews on BPA were funded by the APC and found no consistency of effects was, for vom Saal, more than mere coincidence. He was increasingly vocal about his absolute distrust of and disrespect for the chemical industry. One of his first jabs at the industry came with the 2005 publication of a strong critique of the Harvard Center's weight-of-the-evidence report, which he coauthored with Claude Hughes, an original member of the Harvard Center review committee.[135]

In their paper, vom Saal and Hughes drew attention to a strong relationship between the source of funding and the outcome observed in the published literature on BPA. Since 1997, vom Saal and Hughes reported, 115 studies had been published on low-dose effects of BPA. The researchers sorted these studies by source of funding. Government funding supported 104 of the 115 studies, with industry funding the remaining 11. Of the 104 government-supported studies, 94, or a little over 90 percent, reported significant effects of exposure. Of those 94 government-funded studies, they found that 31 studies reported effects below the EPA's reference dose for BPA. In contrast, none of the 11 industry-funded studies found any effects of low-dose exposure.[136] To be fair, vom Saal and Hughes did not simply conclude that the significant difference in findings was due to a funding effect alone but also considered the possibility of differences in study design and pressures to publish positive results in peer-reviewed literature as sources of bias.

In the article, they cited an extensive list of reported effects of low-dose exposure to BPA, including increased postnatal growth; early onset of sexual maturation in females exposed in utero; increased prostate size in male offspring; decreased sperm production and fertility in males from developmental or adult exposure; stimulation of mammary gland development in female offspring; significant disruption of alignment of chromosomes in developing oocytes; increase in mortality of embryos; disruption of adult estrus cycles; altered immune function; changes in the brain (including increases in the progesterone receptor mRNA, ER alpha levels, ER beta mRNA levels, and somatostatin receptors); behavioral effects (including hyperactivity, increased aggressiveness, and altered reactivity to painful or fear-provoking stimuli); and changes in sexual differentiation of the brain.

This laundry list of critical physiological changes studied by a diverse number of laboratories, vom Saal and Hughes contended, meant that it was high time for a new risk assessment of BPA.[137] Their assessment was not an evaluation of the weight of the evidence but rather an inference of concern based on numerous observations from multiple laboratories. Here was a chemical that interacted with numerous developmental pathways at very low doses, yet was declared safe by its producers and the FDA, a decision based largely on data from industry-funded studies. Regulatory toxicity studies might find no effects of BPA, but the implication was that these studies weren't asking the relevant questions about potential risks at low doses. For the producers facing growing public concern, there was every reason to continue to defend their research and nitpick any data that might call into question BPA's safety.

Only a few years after the NTP called for more research on BPA, over a hundred studies had been published in the peer-reviewed literature. The scope of the effects examined, as noted by vom Saal and Hughes, extended well beyond the original debate about fetal prostate development to include the female reproductive system, the brain, the immune system, and the metabolic system. For example, Ana Soto's lab at Tufts University investigated the effects of extremely low doses of BPA on development of the mammary gland as part of its work on the fetal origins of breast cancer.[138] Cancer researchers at the University of Illinois, led by Gail Prins, studied the association between BPA and precancerous prostate development.[139] A number of replication studies of Pat Hunt's work on BPA's chromosomal effects on female mice eggs were undertaken.[140] As more researchers examined the different effects of BPA, the endpoints measured, the species and strains used, the mechanisms of action identified, and the methods used to deliver doses, the published literature grew more complex and diverse.

Research on the cellular and developmental effects of low-level BPA exposure in laboratory animals introduced a number of complicated questions about the interpretation of such effects. Most important, which effects might be benign perturbations and which might lead to serious health effects? In other words, what constituted an adverse effect? Could low doses have large, long-term effects, as vom Saal and his colleagues argued? Or were humans resilient to low levels of endocrine-active compounds? Whose science counted in determining an acceptable risk? Which study methods were reliable? Which data were relevant? Despite the weight-of-the-evidence assessments that declared there were no human risks of exposure and the FDA's consistent support of BPA's safety, expanding media coverage of the research on BPA was generating a spiraling crisis.

Evaluating or synthesizing animal research on low-dose effects, however, was only one-half of the scientific debate. A contentious debate also revolved around determining how and at which concentrations humans were exposed to BPA. Metabolism studies of BPA reported that the compound was quickly processed in the adult body and excreted in the urine.[141] In 2005, the CDC released the first population-based study of human exposure to BPA. Over 90 percent of the Americans sampled in the study—a representative subgroup of the U.S. population—had detectable levels in their urine.[142] Americans appeared to be getting a steady dosing of BPA every day at very low concentrations. This raised two critical questions. First, how were people being exposed? Second, were the levels present in the population a significant health risk, and, if so, which segment of the population was at greatest risk?

Initially, the two conspicuous sources of exposure were epoxy resins, used to line food cans, and reusable polycarbonate food and beverage containers. (More recent concerns have been raised about exposure through the skin from thermal receipts.)[143] The FDA regulates both sources of oral exposure to BPA as an indirect food additive or food contact substance. By 2005, after a half a century of "safe" use in food packaging and a steady increase in production that topped six billion pounds a year globally, BPA's safety was under considerable scrutiny.[144] Environmental and consumer-based organizations jumped into this debate with greater force and organizing energy and launched product campaigns targeting baby bottles and food cans as sources of BPA.

Advocates used biomonitoring as an organizing tool, testing individuals and products, in order to raise consumer awareness about everyday exposures to hazardous chemicals. Detecting potentially harmful chemicals inside the body raised the moral issue of chemical trespass: "I didn't choose to ingest this." This moral argument sidestepped the more technical debate about metabolism and whether the levels present in the body could have an adverse effect. Detecting the presence of the chemical in products agitated retailers and end users of canned foods and plastic bottles to seek out potential alternatives to BPA. For companies selling baby bottles, in particular, the presence of BPA became a public and potential fiscal liability. Ultimately, the combination of an outraged and fearful public and anxious retailers provided political pressure for legislative reforms at the state and international level.

With the introduction of product campaigns and market pressure, media coverage of BPA increased, and the debate took a decidedly contentious turn. BPA became a poster child for the unchecked dangers of chemicals and government failure to protect the public's health. For this to be a successful political strategy, the certainty of BPA's risks had to be confirmed. From the industry's perspective, the opposite held true: its safety had to be absolute. This political need for certainty provided little conceptual space for the complex challenge of evaluating low-dose risks and interpreting the evidence for decision making—the arena of policy making. One simple question shaped the debate: Was BPA safe or not?

6 Battles over Bisphenol A

In the first decade of the twenty-first century, the question of BPA's safety gained widespread domestic and international attention. The prevailing public narrative about BPA—articulated by advocacy organizations and a growing number of researchers, and echoed by many journalists and social media bloggers—held that it was unsafe. Environmental health advocates across the United States targeted plastic baby bottles and canned foods as unacceptable sources of BPA exposure, aggravating consumer concerns and demands for safer products. Wal-Mart responded by pulling polycarbonate plastic baby bottles off shelves, replacing them with readily available BPA-free bottles—made of metal and other plastics—at premium prices.[1] Sunoco, a BPA producer, adopted a company policy prohibiting sale of the chemical for use in food or beverage containers intended for children three years and younger.[2] Across the United States, opportunistic tort lawyers filed cases against baby bottle and water bottle manufacturers using plastics containing BPA.[3] State bills to ban BPA in baby bottles proliferated in the second half of the decade as advocates turned to legislators to regulate the controversial chemical, whose safety the FDA and the major trade associations continued to firmly uphold.

If BPA is unsafe, as some state governments asserted when banning it from baby bottles, the implication is that the regulatory system failed to adequately protect public health. That legislators and markets responded to public concern might imply regulatory failure, but deciphering whether this was due to inadequate legal authority or capital (financial or human), industry capture of the regulatory process, or the cognitive dissonance generated by new evidence is not an easy task. Moreover, from the perspective of the FDA, maintaining that BPA was safe was not a failure at all. From the agency's perspective, the chemical had not slipped through the cracks of regulation but had been well studied in standardized, validated testing systems and determined to present minimal risks at low doses. Advocates who pointed to new research on low-dose effects and called for product bans, however, used the case of BPA to highlight larger failings of the overall U.S. regulatory system to protect the public from hazardous chemicals. Specifically, they called for federal reform of the 1976 Toxic Substances Control Act (TSCA) and the 1958 Food, Drug and Cosmetics Act. In this case, regulatory failure was framed as a weakness in legal authority and human capital in the agencies.

What became clear in the struggle to challenge or defend BPA's safety was that regulatory success or failure depended on one's viewpoint. For industry representatives, the FDA's consistent reliance on industry-funded toxicity tests to determine a safety standard upheld "sound science." Capture of the agency in this form was achieved through a mutually developed process that involved the use of standardized toxicity tests to determine safety standards. The information generated was narrowly and consistently defined, which allowed for a controlled definition of safety. Research models and testing systems might change, as was the nature of scientific pursuit, but the process to evaluate toxicity and thus define safety could remain relatively fixed. Academic researchers such as vom Saal, Prins, Soto, and Hunt, who observed significant biological effects of low doses of BPA and then saw their work dismissed by the FDA and other international regulatory agencies as inadequate for determining safety, cried foul. The FDA not only had been captured by industry's methods and science, but as a result was unable to rapidly adjust to and integrate new research and evaluation tools and test methods. This dispute over the adequacy or failure of the regulatory system emerged at a time when the country as a whole was growing increasingly frustrated with the federal government.

The inaccurate intelligence assessments used to justify the invasion of Iraq, the mismanagement of the Iraq occupation, and the fatally slow and inept response to Hurricane Katrina's 2005 assault on New Orleans all contributed to the precipitous decline in George W. Bush's approval ratings in his second administration. In the 2006 midterm elections, Democrats successfully took advantage of public dissatisfaction with the ruling party and gained majority control in the House of Representatives. In the presidential election two years later, the voting public narrowly swung to the left. President Barack Obama and the Democratic Party, which gained control of not only the White House but also both houses of Congress, rode into Washington in 2009 with the promise of "change" and major reforms. But the deepening financial crisis, growing political intransigence marked by the rise of the Tea Party, and the stalled economy quickly put progressive reforms on the back burner.

A fierce public relations battle raged between advocates and industry in contests over regulating BPA. The scientific and political warfare described in this chapter has implications that go well beyond the future of this one chemical. The struggle is not just about whether BPA should or should not be banned from certain products. A more fundamental challenge revealed by this dispute is how emerging research on chemical risks can effectively improve public health and product safety. Banning products and chemicals one by one will never achieve this goal. The question is whether political attention generated by campaigns for chemical bans in general is sufficient or effective in meeting this challenge. How will evidence on the health effects of chemical exposure be used to make effective public health decisions and reduce risks? This cannot be solved simply through sound science practices that seek to bind the definition of a chemical's safety to traditional toxicity testing and a priori assumptions about the dose-response relationship. As the United States and the rest of the world face a growing burden of chronic disease, which questions need to be asked about the risks of chemical exposure? Can environmental health research more directly inform health policies for disease prevention? There is no one-word answer to these questions. Nor is the answer to the question "Is BPA safe?" a simple yes or no. But the solutions do not rest in the status quo. The old paradigm—mitigating risk by assuming safety at low doses and defining safety in relation to toxicity or even carcinogenicity measured at high doses—no longer fits reality today. Humans are constantly exposed to hundreds of industrial chemicals, from conception to old age, and researchers are developing tools to investigate the biological impacts of such exposure. It is essential that new systems to evaluate chemical risks emerge.

The proliferation and volume of research driven by new tools and technologies, such as rapid screens to detect gene-chemical interaction, as well as new understandings of gene-environment interactions, or epigenetics, have outpaced our capacity to interpret new discoveries. There are lots of dots, but no agreement on how to connect them, and still less agreement on what to do about the connections that are made. This cognitive dissonance provides ample opportunity to manufacture doubt, but it also generates demand for new tools and methods for independent, transparent, and democratic assessment of the evidence for decision making. As Tina Bahadori, the former director of the American Chemistry Council's (ACC's) Long-

Range Research Initiative, who joined the EPA in 2012, cogently explained the problem: "These technologies demonstrate changes . . . without knowledge of what those changes mean (minor compensation or adversity) or how they relate to the exposure and doses people would actually receive. Without investment in the science of *interpretation*, the tendency will be to rely on high-through-put hazard data or biomonitoring as a surrogate for risk assessment."[4] Who does this interpretation reflects how democratically information will be used to inform decisions about safety and risk.

Interpretation of science contains its own set of difficult and portentous questions. Which data and whose data should be used to evaluate risk—a critical focus of sound science? Which endpoints should be measured? Industry argued forcefully during the Endocrine Disruptor Screening and Testing Advisory Committee (EDSTAC) debates and to the National Toxicology Program (NTP) (see the preceding chapter) that traditional toxicological endpoints should continue to be used to evaluate low-dose effects of hormonally active agents. What assumptions were made about dose-response relationships? The Annapolis Center's paper on toxicology called for the maintenance of a fundamental principle: the dose makes the poison. Meanwhile, vom Saal and Gorski explained hormones' nonmonotonic dose-responses in the National Academy of Sciences' report on hormonally active agents.[5] Which effects indicate disease? What is the burden of proof for action: certainty of human causation of disease, or a critical level of concern? These are the questions about the validity and reliability of data that underlie the contemporary debate about BPA. The argument focuses on the interpretation of BPA data, which explains why, during the early 2000s, different groups of researchers came to different conclusions about the same research.

The story told thus far in this book does not lead to a simple ending or solution, whether a call to uphold the precautionary principle or a demand for radical reform of the regulatory process or chemical legislation. Instead, it offers a more modest recommendation for enhancing democratic decision making amid growing scientific complexity, and given inherent inequities in the distribution of political power that shape the process of interpreting science. My recommendations include the development of a more transparent and independent process to evaluate diverse streams of evidence and develop criteria used in decision making. Transparency, in this case, does not simply mean data disclosure. Rather, as discussed at the end of the chapter, transparency requires that values, biases, conflicts of interests, and assumptions be made explicit to render the interpretative process more open. Formalizing the process of data interpretation and decision making—what we know and how to use that information—is demanded by industry producers for market stability, to strengthen consumer confidence in safety, to improve the public's trust in the regulatory decision-making process, to shift chemical production toward safer production, and, hopefully, to improve public health.

BPA RISKS: LOW DOSES, EPIGENETICS, AND FETAL ORIGINS OF DISEASE

Over two unseasonably warm days in late November 2006, the National Institute of Environmental Health Sciences (NIEHS) hosted an expert panel in Chapel Hill, North Carolina, to evaluate the state of the science on BPA. Organized by Jerrold ("Jerry") Heindel from the extramural research, or grant-making, arm of the NIEHS, the meeting included a little over three dozen government and university researchers from the United States, Europe, and Japan, all of whom were directly working with BPA or within the field of endocrine disruption. Among them were John McLachlan from Tulane University; Antonia Calafat, the principal investigator on biomonitoring studies conducted by the CDC; and Linda Birnbaum, a leading toxicologist at the EPA, who was appointed NIEHS director in 2009 under President Obama. Many of the U.S. researchers were recipients of research grants from the NIEHS under a program Heindel initiated in 2002, "The Fetal Basis of Adult Disease: Role of the Environment."[6]

Heindel is a reproductive biologist by training, with expertise in the function of the Sertoli cell, a specific cell in the male testes that "nurses" developing sperm. A straight-shooting researcher turned administrator, he joined the NIEHS in 1987, at the time bringing a new biological perspective to the reproductive toxicology program, which was then dominated by toxicologists. Heindel's arrival was part of a larger effort by the NIEHS director, David Rall, to foster greater disciplinary diversity within the field of environmental health science. For close to ten years, Heindel followed emerging research on endocrine disruption. Strongly skeptical at first, he became convinced, as the research piled up over time, that there was something to the theory. It was clear that chemicals were capable of interfering with endocrine function or signaling during critical developmental periods. Yet, as Heindel noted, there was little research that followed exposed animals for long periods of time to determine whether these endocrine alterations or disruptions manifested in increased disease risk.[7]

Many of the new studies on BPA and other endocrine disruptors in the early 2000s focused on identifying new mechanisms of action, such as binding to different receptors or turning on or off various genes. The focus was on the cell. Rapid screens to detect the presence of genes, DNA libraries, and microarrays to identify genes and study their function and interactions all generated new ways to study the relationship between genes and disease. The revelation that 99.9 percent of the human genome is identical across the world's population crushed the expectation that sequencing the human genome would unlock the mysteries of human life and reveal magic bullets for curing disease. Except in rare cases, disease does not arise simply as a result of the presence of a single gene but may be better understood through the "ecology of gene expression."[8] In other words, the presence of certain genes may be necessary but not sufficient for disease development.

In the "postgenomic age" of the early twenty-first century, researchers turned their attention to the products of genes—proteins, enzymes, metabolites—and their function, as well as to patterns, or ecology, of gene expression. Which changes or patterns of changes (e.g., enzymes, hormones, proteins, drugs, nutrition, stress) in the environment might cause different patterns of genetic expression that lead to abnormal development or function, disease, or increased disease susceptibility? Causal models for disease development were mapping increasingly complex, heterogeneous interplays between the environment and genes. The study of BPA became enveloped in this larger pursuit as researchers examined the compound's interaction with the human genome and the newly understood epigenome.

The study of epigenetics seeks to elucidate the relationship between the environment and genes. Prior to identification of the human genome, epigenetics, very broadly, meant phenotypic expressions not directly explicable by genetic change. For example, in the 1970s, "epigenetics" in cancer research referred to any carcinogenic activity not resulting from gene mutation. Estrogens, for example, were potential carcinogens that did not act through genetic mutation. By the early 1990s, epigenetics was more narrowly specified as the study of the epigenome, chemicals that lie along the double helix strands of DNA and turn gene expression on or off. Changes in the epigenome due to exposure to industrial chemicals, vitamin deficiency, or stress may alter gene expression and function.[9]

Epigenetics provides greater mechanistic grounding for understanding the effects of endocrine-disrupting chemicals. McLachlan, who led the study of environmental estrogens, and in particular DES, and his colleague David Crews, an evolutionary biologist, argue that endocrine disruptors interact with genes and the epigenome as well as communication between the two, affecting biological development and function. These chemicals "do not act on genes alone but on developmental mechanisms that integrate genetics and epigenetic interactions, resulting in the phenotype."[10] McLachlan and Crews suggest that epigenetic effects can help to explain the transgenerational effects of DES observed in animals three decades ago. Part of the explanation is the fact that early in development, the "ability of the genotype to produce different phenotypes" (i.e., its plasticity) is at its greatest.[11] Therefore, exposure during this very early period can have long-term and multigenerational effects on an organism's phenotype. A 2005 study published in *Science* demonstrated powerful epigenetic effects in rats exposed to two endocrine-disrupting compounds, vinclozolin and methoxychlor, early in development. Single exposures to one of these compounds during pregnancy resulted in significant effects on sperm volume and viability in the next four generations of (unexposed) male rats.[12] In other words, a single exposure in one generation was showed to alter, through epigenetic mechanisms, reproductive development in later generations of rats never exposed to the chemical itself. Epigenetic effects, then, might further explain how environmental exposures—including nutrition, stress, maternal smoking, diabetes, and synthetic chemicals—during fetal development might led to disease in adulthood.

For Heindel, such advances in epigenetics research lent powerful support to theories of the fetal origins or basis of adult disease and were potentially a breakthrough insight into how endocrine-disrupting chemicals might increase disease risk.[13] In 2002, he launched a two-million-dollar research grants program at the NIEHS to expand investigations into the long-term disease risks of in utero exposure to endocrine-disrupting chemicals. Heindel made it very clear that proposed research needed to have direct public health relevance. This meant that the chemicals chosen needed to be those found in pregnant women and had to be tested at levels relevant to human exposure, and the studies needed to investigate diseases or abnormalities of public health importance. These included a number of chronic diseases and conditions of rising incidence in the U.S. population, including reproductive cancers, obesity, diabetes, polycystic ovary syndrome, hypertension, neurodegenerative disease, and a host of reproductive problems.[14] In many ways, Heindel directed researchers to make their work increasingly relevant to public health risks. Researchers might spend a lifetime discovering new ways in which chemicals interact with the cell, but such work could not provide the information necessary to reduce potential risks of these exposures to humans.

Among the researchers funded through the program were several studying BPA, including Ana Soto at Tufts University, who was working on the development of the mammary gland and breast cancer, and Gail Prins at the University of Illinois–Chicago's Department of Urology, who was studying effects on prostate gland development and cancer. Like the Tufts team, which, for decades, had studied the role of estrogens in the development and function of the mammary gland, Prins's lab explored hormonal control of the prostate gland's development and function and neonatal exposures to estrogen and effects on the prostate—what is called "developmental estrogenization."[15]

Prins first learned about BPA when she served on the review panel for the NTP's 2000 low-dose report on environmental estrogens as an expert in hormonal control of prostate development. She had followed reports of environmental estrogens for several years, but it was not until after her participation in the NTP panel that she decided to incorporate BPA into her own research. She was at first quite skeptical that BPA would have any effect on prostate development, but when her lab saw hormonal imprinting and later epigenetic changes as a result of low-dose exposures, she decided to pursue the research.[16]

Hormonal imprinting involves the first interactions between a hormone and its receptor at critical periods of development, which inform the signaling or communication capacity of the cell. This capacity influences the organization and development of tissues and organs. Under the NIEHS grant from Heindel's program, Prins's lab began reporting increases in precancerous lesions in the prostates of animals exposed early in development to low doses of BPA and provided a possible explanatory epigenetic mechanism of action.[17]

As early results came in toward the end of the grant period, Heindel decided to organize the Chapel Hill meeting of BPA researchers to review the emerging evidence. Heindel modeled the meeting on the consensus workshops on endocrine disruption, such as the Wingspread Statements (see chapter 4). The objective of the group was not to determine what might be a "no observable adverse effect level" (NOAEL) or "lowest observed adverse effect level" (LOAEL) to use in determining a safety standard, or to decide whether BPA causes harm in humans. Rather, the goal was to determine where there was the greatest certainty of knowledge, given all the existing research, and where there was less certainty. In other words, what was known on the basis of existing data and according to researchers with specific expertise in the field?

For three days, researchers considered a series of questions designed to induce them to integrate information presented in literature reviews precirculated to the group and drafted by selected participants with identified expertise. For example, different groups of researchers were asked to consider the extent to which the low doses used in laboratory animal studies were comparable to the levels detected in human tissues, serum, and urine and the degree to which in vitro evidence provided mechanisms that explained or predicted in vivo studies.[18] Groups were first divided into areas of expertise—for example, wildlife research, human exposure, cancer, genetic and epigenetic research—and then individuals were randomly sorted into groups to determine consistency in responses.

In the months after the meeting, drafts of a single report on the group's consensus were circulated, edited, and prepared for publication. The final report included a series of statements divided into three categories that ranged in their degree of certainty: confidence, or statements on which all groups came to absolute agreement; statements that were "likely but require confirmation"; and "areas of uncertainty and suggestions for future research." All statements addressed four issue areas: in vitro mechanisms, wildlife effects, laboratory animal research and human exposure, and life-stage effects.[19]

Consensus at the highest level of certainty focused on a number of critical points. First, the group outlined its certainty that there is considerable complexity in the mechanistic activities of BPA. Yes, BPA binds to estrogen nuclear receptors, but it also binds to estrogen receptors on the cell membrane, which are more sensitive to BPA, as well as to androgen and thyroid hormone receptors. When conducting research, therefore, one should not simply assume a single mechanism of action, use a narrow range of doses, or presume that similar effects will be observed in different tissues at the same dose.[20] This conclusion conflicted directly with weight-of-the-evidence reviews' assumption of a single mechanism of action.

As for reported effects, the group declared it was confident that BPA "alters 'epigenetic programming' of genes" in animals in ways that can result in persistent effects later in life. These effects, or what the group called "organizational changes," include changes in the development of the prostate, breast, testes, mammary glands, and brain. It went on to conclude with confidence that these effects vary with the life stage at which exposure occurs because there are developmental windows of susceptibility for different tissues.[21] In other words, the long-term effect observed depends not only on the exposure concentration but also on when that exposure occurs in development.

The thirty-eight researchers who put their names to the BPA consensus report contended that the certainty of human exposure, organizational changes, and epigenetic programming in multiple systems in laboratory animals raised the level of concern for human health effects. While the report did not claim that organizational changes observed in studies necessarily resulted in adverse events, this understanding did not mean they were certain of no adverse effects—quite the opposite. The certainty of effects observed gave reason for caution and concern about long-term adverse effects. Here the process diverged considerably from the Annapolis Center's weight-of-the-evidence approach, which set causation of human harm as its evaluative standard. Below this standard, there was no concern. It was all or nothing, a bright line drawn at causation. In contrast, the consensus report attempted to delineate different levels of uncertainty to reveal what was known.

Two very different approaches by two different groups of people came to two different conclusions regarding the risks of BPA. The industry-sponsored weight-of-the-evidence reviews declared that BPA does not harm humans, a conclusion based primarily on two large regulatory toxicity tests that specifically examined well-documented and accepted adverse toxicological effects and reported no effects, and on the lack of consistency among specific adverse effects across smaller published studies. In contrast, the consensus statement concluded with certainty that widespread human exposure, together with evidence of organizational change in animal models, provided a basis for concern about long-term health effects in humans. Caution was warranted.

This difference in the interpretation of the same scientific evidence generated public confusion and debate not only about the safety of this one chemical but more broadly about whose science counts and who is a legitimate interpreter of evidence. Industry representatives argued time and time again that the large studies they sponsored or conducted, which followed good laboratory practices (GLP) and examined well-defined adverse endpoints across a range of dosing groups (and used oral dosing rather than subcutaneous injections), represented more rigorous research—the studies were considered both relevant and reliable for regulatory decision making. Further, industry representatives pointed to the limitations and flaws of academic research: the smaller number of animals, the fewer dosing groups, the use of nonoral routes of exposures.[22]

The economic influence on this epistemic approach to evaluate chemicals cannot be overlooked. It is not in the immediate interests of BPA producers to explore the uncertainties and biological complexities of the chemical's interaction with hormone receptors, developing tissues, cellular communication, and gene expression. To conclude, as the Chapel Hill consensus report did, that given the known complexities it is difficult to state with complete certainty that BPA does not inflict harm on the developing organism would open the door to new considerations of risk and expand the meaning of harm, with obvious negative impacts on the chemical's market. For an industry producer to take such a position would be potentially self-defeating.

For environmental advocates, however, the conclusions of the Chapel Hill consensus statement were clear and immediately championed: BPA presented a risk to public health. Without an immediate economic stake in BPA's market and already sympathetic to the belief that the chemical was hazardous, advocates were willing to call the chemical unsafe. "Organizational changes" and evidence of exposure were interpreted as sufficient proof to support product bans and raise public concern about the chemical's safety. The independence of the consensus process from industry representation—not a single representative was present at the meeting—lent further credibility to the report, particularly for the advocacy community.

Whether BPA was safe or not increasingly became a stark, black or white issue dividing public interest organizations and industry. This is a contemporary iteration of the David-versus-Goliath, industry-versus-the-people narrative that runs throughout histories of battles over chemical pollution.[23] Campaigns launched against Rachel Carson by the agricultural industry after the publication of *Silent Spring* in the early 1960s, and against Herbert Needleman by the lead industry in the 1980s, give evidence of the intensity and sometimes vicious nature of the battle waged against individuals who generate considerable public attention to serious risks of chemicals.[24] By 2006, as the debate narrowed around whether the compound was safe and whether a ban was warranted, the political stakes had been raised and the battle lines drawn.

WHOSE SCIENCE IS SOUND?

In response to the proliferation of research on BPA and growing public concern about its safety, the NTP's Center for the Evaluation of Risks to Human Reproduction (CERHR) decided to initiate a review of the scientific literature on BPA in late 2006 and early 2007. Established in 1998, the CERHR reviewed scientific research on chemicals that might have reproductive or developmental effects for use in risk assessments by regulatory agencies. The organizers immediately found themselves thrust into the middle of a political fight. BPA producers, researchers, and environmental advocates appeared at every public meeting, along with television crews and reporters. Members of Congress were even called upon to intervene.

Before the CERHR external scientific panel, chosen to review the literature on BPA, could even meet, advocates cried foul play. It turned out that CERHR had contracted with the private firm Sciences International to draft a review of the literature that would be used as the basis for the panel's review. This was not out of the ordinary. Since its creation in 1998, CERHR had held contracts with Sciences International to draft scientific reviews.[25] This time, however, the advocacy community was intently focused on every step of the review process. In the days leading up to the panel's first public meeting, the *Los Angeles Times* ran a front-page article that disclosed a potential conflict of interest in the draft report: the article alleged that Sciences International also held contracts with Dow, a major BPA producer.[26] An investigative report in the magazine *Fast Company*, published two years later, dug a little deeper and alleged that in addition to conducting scientific reviews for federal institutions and private companies, Sciences International represented private firms in lawsuits and provided assistance in navigating the regulatory process. According to the article, the company lobbied on behalf of R. J. Reynolds in the late 1990s to block EPA efforts to establish stricter standards for the pesticide phosphine that included an expedited review of the science by Sciences International.[27]

Disclosure of its contracts with chemical firms, together with the combination of legal, regulatory, and scientific skills housed at Sciences International, raised serious questions about whether the contractor was part of the lucrative product defense industry. Representative Henry Waxman (D-CA) and Senator Barbara Boxer (D-CA) sent letters to David Schwartz, director of the NIEHS, requesting a briefing on the matter. At the time, Schwartz faced a separate congressional investigation of his potential mismanagement, fiscal improprieties, and political influence at the NIEHS, which included alleged efforts to privatize the institute's journal *Environmental Health Perspectives*. Persistent congressional pressure resulted in his resignation in late 2007.[28] The NIEHS quickly responded to Waxman and Boxer's inquiry into Sciences International by canceling the contract and initiating an audit of the contractor's report on BPA.[29]

Scrutiny by environmental advocates did not stop with Sciences International. Advocates from organizations such as the Natural Resources Defense Council and Environmental Working Group also publicly raised questions and concerns about the selection of the experts serving on the CERHR panel. While panel members were selected via an open nomination process, advocates questioned why none of the more outspoken researchers working and writing on BPA, such as vom Saal, Soto, Prins, and Hunt, who participated in the Chapel Hill meeting, had been asked to serve.[30]

Determining who serves on an evaluative panel is often a contentious process, given that the outcome of the evaluation may be ascribed to the personalities in the room—the weight on the panel.[31] Removing or selecting specific individuals because of their known position on an issue can, subsequently, strongly influence the outcome of a panel's findings. For example, in 2007, after a written request from the ACC, the EPA dismissed Deborah Rice, a toxicologist from the Maine Department of Health and Human Services (and formerly with the regulatory agency Health Canada and the EPA), as chair of an expert panel evaluating the toxicity of a flame retardant and removed all of her comments from a draft assessment. Rice held no financial conflicts of interest but had publicly expressed serious concern about the safety of the chemical based on her research and knowledge of the literature. However, the EPA argued that Rice's position on the issue of toxicity presented a conflict of interest and disqualified her from the panel.[32]

In selecting the panel participants, the CERHR staff sought to manage the heated politics of the BPA debate by excluding researchers with considerable experience in studying the chemical. Holding expertise and explicit values became conflated with bias and conflict of interest. Since researchers such as Soto and vom Saal had taken strong positions on the risks of BPA, they were not considered to be objective participants. However, the decision to keep vom Saal off the panel might also have been personal. In a number of scientific meetings, vom Saal had come to verbal blows with various industry researchers as well as with L. Earl Gray of the EPA. Gray had conducted extensive work on phthalates and several studies on BPA, and sat on the CERHR panel.

While conflicts of interests, values, and biases can be managed, eliminating them entirely is not possible. Yet this is exactly what CERHR attempted to do; in an effort to eliminate perceived value differences and personal bias and to avoid uncomfortable personal conflicts and confrontations, the panel lost considerable expertise on BPA. At the same time, CERHR fell short in managing conflict of interest by failing to account for the financial conflicts of its contractor, Sciences International. Clearly, a well-defined policy or process to manage conflict of interests and balance inevitable scientific debate and bias was lacking. This weakness became all the more evident in the audit of the Sciences International contract.

The audit involved a simple counting of the number of recommendations made by the CERHR panel that were incorporated by the contractor during the drafting of the initial report, and the number of published studies on BPA found in the open literature compared to the number of studies included in the contractor's draft report. On the basis of these two considerations, the NIEHS concluded there was no conflict of interest, and the original report prepared by Sciences International for CERHR was retained.[33] What the audit did not attempt to evaluate was the interpretative process, or even communication between industry producers and the contractor. Which assumptions were used in evaluating the research? Which criteria were used to evaluate the bias, validity, or relevance of the studies, and why?

Accepting the Sciences International draft as valid, the CERHR panel applied five criteria to evaluate what it called the utility of a study. (The criteria were the route of exposure, with greater weight given to oral exposure; appropriate exposure design and statistical analysis; adequate sample size; proper use of the material employed to expose test animals; and the use of a positive control.) On the basis of the quality of the studies as determined by these criteria, the panel then evaluated the strength of evidence for various endpoints measured. One critical outcome of using these criteria was the marginalization of much of the low-dose research, such as Prins's and Soto's research on prostate and mammary gland effects, due to the fact that animals were not exposed to BPA orally. Prins and Soto argued that exposing animals via injections or internal pumps provided them with greater control over the internally delivered dose, particularly given the very low concentrations they used. The debate over the appropriate route of exposure continues to be an important variable in the evaluation of BPA safety today. This is because oral exposure, which is the presumed dominant route of exposure, results in a metabolic process in the gut and liver that converts over 90 percent of the chemical to a form that is not actively estrogenic (i.e., BPA glucuronide). This detoxification process is bypassed when BPA is released directly into the blood of an animal (by, e.g., injection or internal pumps). The focus on oral dosing, as opposed to the internal dose, assumes that the only route of exposure of concern is food. This assumption began to be questioned more seriously as researchers raised concerns about the presence of BPA in thermal receipts at levels a thousandfold higher than levels detected in food.[34] Disagreements about the utility of nonoral routes of exposure led Prins to measure the internal concentration of unconjugated BPA or the biologically active form after oral exposure and injection. She later reported that the levels of active BPA were similar after several hours regardless of the method of dosing.[35]

The panel organized its findings into five categories of concern, ranging from the lowest, "negligible," to the highest, "serious concern," with "some concern" falling in the middle. The first draft of the CERHR's report concluded it had "some concern" that low-level BPA exposure was associated with neural and behavioral effects, but "minimal concern" for prostate effects and accelerated puberty.[36] Most of the low-dose work was deemed of limited utility, and researchers from Soto's and vom Saal's labs reacted to the panels' conclusions with frustration and extensive criticisms.[37]

Industry officials greeted the first draft of the report with considerable praise, heralding it as a victory for sound science. In a press release on the panel's findings, Steve Hentges, a spokesperson for the ACC, stated that the CERHR had confirmed that BPA presented "minimal concern" for human health, choosing to ignore its statement about "some concern." "Most importantly," Hentges argued, "these conclusions are from a very credible, highly qualified group of independent scientists with no conflicts of interest, operating in an open and transparent review process." He went on to disparage the Chapel Hill review as a process that was "not open and transparent" and lacked a conflict-of-interest policy.[38]

The final report, or monograph, issued by the NTP nearly a year later increased its level of concern about BPA's prostate risks. While confirming the panel's conclusion of "some concern" for effects on the brain and behavior "in fetuses, infants and children at current human exposures to bisphenol A," the report raised the level of concern for prostate effects from "minimal" to "some concern."[39] The different conclusion about prostate effects was based on the NTP's acceptance of low-dose studies that exposed animals nonorally, as well as on new data on subtle cellular effects not available to the CERHR panel.[40] During preparation of the final monograph, NTP staff raised the level of concern for mammary gland effects from "minimal" to "some concern," again because of consideration of nonoral studies. This decision was later revised back to "minimal concern," on the basis of the recommendations of a pathologist who concluded that the reported mammary gland effects were mild and not necessarily the form of greatest concern in the development of invasive breast cancer in women.[41]

The NTP report had an immediate and considerable impact on the public debate about BPA safety. It provided advocates with an authoritative position and statement on the chemical's risks to take to state legislators considering product bans, and it influenced the ongoing Canadian assessment of BPA. In Canada, the environmental advocacy group Environmental Defence (independent of the U.S.-based Environmental Defense Fund) raised the debate about BPA risks to the national level. The organization's

Sponsor	Title	Institution	Date Released	Key Findings
National Toxicology Program	Carcinogenesis Bioassay of Bisphenol A (CAS No. 80-0507) in F344 Rats and B6c3fl Mice (Feed Study)	Litton Bionetics	1982	"[N]o convincing evidence of carcinogenicity"; "that 'bisphenol A is not carcinogenic' should be qualified to reflect the facts that leukemia in male rats showed a significant positive trend, that leukemia in high-dose male rats was considered not significant only on the basis on the Bonferroni criteria, that leukemia incidence was also elevated in female rats and male mice, and that the significance of interstitial-cell tumors of the testes in rats was dismissed on the basis on historical control data."[26(p8)]
National Institute of Environmental Health Sciences/ Environmental Protection Agency	National Toxicology Program's Report of the Endocrine Disruptor's Low-Dose Peer Review	National Toxicology Program	2001	"There is credible evidence that low doses of BPA [bisphenol A] can cause effects on specific endpoints. However, due to the inability of other credible studies in several different laboratories to observe low dose effects of BPA, and the consistency of these negative studies, the Subpanel is not persuaded that a low dose effect of BPA has been conclusively established as a general or reproducible finding."[50(pvii)]
American Plastics Council	Weight of the Evidence Evaluation of Low-Dose Reproductive and Developmental Effects of Bisphenol A	Harvard Center for Risk Analysis	2004	"The panel found no consistent affirmative evidence of low-dose BPA effects for any endpoint. Inconsistent responses across rodent species and strain made generalizability of low-dose BPA effects questionable. Lack of adverse effects in two multiple generation reproductive and developmental studies casts doubt on suggestions of significant physiological or functional impairment."[56(p875)]
American Plastics Council	An Updated Weight of the Evidence Evaluation of Reproductive and Developmental Effects of Low Doses of Bisphenol A	Gradient Corporation	2006	"No effect is marked or consistent across species, doses and time points. Some mouse studies report morphological changes in testes and sperm and some non-oral mouse studies report morphological changes in the female reproductive organ. Owing to lack of first pass metabolism, results from non-oral studies are of limited relevance to oral human exposure."[61(p1)]
National Institute of Environmental Health Sciences/ National Institutes of Health	Chapel Hill Bisphenol A Expert Panel Consensus Statement	National Institute of Environmental Health Sciences and invited BPA experts	2007	"We are confident that . . . human exposure to BPA is variable, and exposure levels cover a broad range [central tendency for unconjugated [active] BPA: 0.3–4.4 ng ml-1(ppb) in tissues and fluids in fetuses, children and adults . . . Sensitivity to endocrine disruptors, including BPA, varies extensively with life stage, indicating that there are specific windows of increased sensitivity at multiple life stages . . . BPA alters "epigenetic progamming" of genes in experimental animals and wildlife that results in persistent effects that are expressed later in life...Specifically, prenatal and/ or neonatal exposure to low doses of BPA results in organizational changes in the prostate, breast, testis, mammary gland, body size, brain structure and chemistry and behavior of laboratory animals."[67(p134)]
National Toxicology Program	NTP-CERHR monograph on the potential human reproductive and developmental effects of bisphenol A	Sciences International, Center for the Evaluation of Risks to Human Reproduction (CERHR)	2008	"[S]ome concern for effects on brain, behavior and prostate gland in fetuses, infants and children at current human exposures to bisphenol A."[68(pvii)] "[S]ome concern for effects on brain, behavior and prostate gland in fetuses, infants and children at current human exposures to bisphenol A."[68(pvii)] "[T]he possibility that bisphenol A may alter human development cannot be dismissed."[68(p7)]

Figure 6. Reviews of bisphenol A conducted in the United States. (From Sarah A. Vogel, "The Politics of Plastics: The Making and Unmaking of Bisphenol A 'Safety,'" *American Journal of Public Health* 99, suppl. 3 (2009): S559–66, American Public Health Association.)

director, Rick Smith, later coauthored a popular book, *Slow Death by Rubber Duck* (2009), on chemicals that, like BPA, leach out of consumer products.[42] BPA became a political football, with the Conservative federal government and Liberal provincial government of Ontario vying to champion the issue. In 2008, Health Canada, the federal public health agency, announced a decision to take "precautionary action to reduce exposure [to BPA] and increase safety" and declared BPA "toxic." Health Canada based its decision on evidence of "some concern" for "brain, behavior, and prostate effects in fetuses, infants and children at current human exposures to bisphenol A," as concluded by the NTP.

This "toxic" category comes directly from the Canadian Environmental Protection Act (CEPA) of 1999, which established a process to prioritize and evaluate the safety of chemicals. While the Canadian evaluation process is principally grounded in risk-based approaches that consider exposure and effects in risk assessment, it also incorporates aspects of the precautionary principle in risk-management recommendations. Importantly, as in Europe, the Canadian system keeps the processes of assessment and management separate and distinct from each other.[43] Therefore, while the Canadian risk assessment of BPA didn't come to a radically different conclusion from the NTP, the risk-management decision allowed for a more precautionary action based on potential risks to the very young. The Canadian government's precautionary decision on BPA, the first in the world, involved regulations restricting the "importation, sale and advertising of polycarbonate baby bottles that contain" BPA, as well as some "action to limit the amount of bisphenol A that is being released into the environment."[44]

CRISES IN CHEMICAL SAFETY

A siege was rising against BPA producers. "We are under attack from all fronts," declared the president of the Society of the Plastics Industry (SPI) to its members at an annual meeting in 2009. Producers considered the battle to be at a major turning point, as legislators and regulators considered action that would "fundamentally change our business model."[45] The release of the Chapel Hill Consensus Statement, the NTP's report of "some concern," state bills to ban BPA, the Canadian government's listing of BPA as "toxic," and the first of a number of liability lawsuits all contributed to a worsening headache for industry.[46]

Environmental advocates, working collaboratively across the United States and Canada, launched a multipronged strategy to restrict the use of BPA that applied pressure on sensitive producers, most notably baby bottle manufacturers, and state legislators. In 2008 Representatives John Dingell (D-MI) and his colleague Bart Stupak (D-MI) announced an investigation into BPA "in products intended for use by infants and children."[47] Dingell and Stupak sent letters to a number of infant formula companies asking "how often BPA is used in such lining [infant formula cans], whether the companies that produce the infant formula are aware that BPA is being used in these manners, and if they have tested their product for the presence of BPA."[48] In addition, advocacy-led petitions issued through shareholder actions, combined with consumer pressure on major retailers, drove suppliers to demand alternatives to BPA-based polycarbonate plastic. As noted at the start of this chapter, Wal-Mart announced that it would remove polycarbonate baby bottles from its shelves in 2009.[49] Baby bottles and reusable sports water bottles with "BPA-free" stickers began to appear on store shelves. As the market in baby bottles shifted, legislative campaigns to ban BPA in baby bottles picked up traction.

The first such ban was introduced in California in 2005 in a bill sponsored by Assembly Member Wilma Chan, a Democratic representative from Oakland, with the support of Environment California and the Breast Cancer Fund. Lacking the NTP's report and significant market pressure, the bill failed to pass. But the bill provided a template that other states could use. California had taken a similar lead when it successfully passed a ban on the flame retardant polybrominated diphenyl ethers, or PBDEs, in 2003; eight other states followed. (As it turns out, pentaBDE, which is in the family of PBDEs, has been largely replaced with chlorinated tris and other mixtures. Chlorinated tris is the flame retardant banned for use in children's sleepwear in 1977 after it tested positively for mutagenicity. It was later found to be a probable human carcinogen.)[50] And indeed, after the California BPA bill, dozens of other states introduced legislation to ban BPA. By the time a California BPA ban passed at last, in 2011, eleven states had already passed some form of a ban.[51]

Media coverage, by both traditional and increasingly social media outlets, of BPA research and bans has grown considerably over the past five years, dispersed by parenting bloggers, local news outlets, and some national reporters. Such widespread coverage has drawn the attention of conservative critics, who decry the reporting as advocacy's capture of the media.[52] Indeed, reporting on low-dose BPA studies has been extensive and, at times, of uneven quality. A simple search of coverage of BPA in 2011 on the website Environmental Health News, a source of daily media

Figure 7.   Rally for bisphenol A ban in Sacramento, California, 2009. (Photo credit: Breast Cancer Fund.)

reports on environmental health research from around the world, drew up close to three thousand articles.[53]

Among the more controversial reporting on BPA was a yearlong investigative series conducted by the *Milwaukee Journal Sentinel* in 2008–9, which garnered national attention and even a PBS special in 2008.[54] The newspaper's investigation focused on the close relationship between chemical trade associations and the FDA, as well as the industry's public relations strategies as it fought state bills to ban BPA.[55] This dedicated attention from the media and the prevailing public narrative that BPA was unsafe contributed to growing anxiety within industry trade groups. The crisis over BPA was spinning out of their control. Media coverage, state bills, retail pressure for alternatives, and growing public anxiety about BPA all suggested that industry was losing its fierce public relations battle to defend the chemical's safety.

In a leaked 2009 memo of meeting minutes of the BPA Joint Trade Association, which at the time included the North American Metal Packaging Alliance, the American Chemistry Council, Coca-Cola, Alcoa, the Grocery Manufacturers Association, and Del Monte, participants discussed their considerable frustration with the seemingly intractable public narrative about the dangers of BPA. They needed to be more "proactive" and aggressive in their communications strategy. According to the minutes, participants "suggested using fear tactics (e.g., 'Do you want to have access to baby food anymore?')." Another proposed tactic involved highlighting the "impact of BPA bans on minorities (Hispanic and African American) and poor" and "befriending people that are able to manipulate the legislative process." Industry was losing the public narrative and desperately needed a multipronged communications strategy and a credible spokesperson. "The [communications] committee doubts obtaining a scientific spokesperson is attainable." Ideally, the "holy grail" spokesperson, the memo outlined, would be "a pregnant young mother who would be willing to speak around the country about the benefits of BPA."[56]

Advocates clearly had gained significant political advantage by framing the risk of BPA around children, focusing on baby bottles and, to a lesser extent, infant formula as sources of exposure. The message was clear: baby bottles are poisoning your infant. A consortium of NGOs released a report in 2008 entitled "Baby's Toxic Bottle," the cover of which shows a picture of a baby drinking milk from a bottle, looking up at the camera at such an angle that the viewer feels as if she were holding the child. The report presented the results of NGO-sponsored tests of various baby bottle brands for BPA along with summaries of major concerns about the chemical and recommendations for industry, government, and concerned consumers.[57]

Demanding safe products for children proved to be a winning political argument regardless of party affiliation. Indeed, many states passed BPA bans with broad support from both Democratic and Republican lawmakers.[58] Fighting fire with fire, industry trade associations allegedly used fear tactics to block bills targeted at children's safety. They argued that store shelves would be empty of canned food and infant formula if BPA-based epoxy resins were restricted from children's foods and containers. In several states, industry trade associations allegedly targeted state Women, Infants, and Children (WIC) programs to build support for blocking bills that, they contended, would negatively impact low-income populations.[59]

Chemical bans long have provided effective organizing tools for advocates, and BPA is clearly the latest example of this fact. Efforts to ban DDT in the late 1960s galvanized a nascent environmental movement, and attempts to ban PVC in the 1970s sought to test the scope of the Delaney clause. What, then, is the larger objective of banning BPA? What larger political narrative is being told? To imply that the BPA ban is itself is the goal suggests that the problem can be narrowly defined by the hazard itself, and, as a corollary, that the solution is as simple as removing the compound. Indeed, retailers which prominently display a "BPA-free" label seek to conflate the absence of the chemical with the certainty of their products' safety. Yet replacing one product with another without reform of how safety is defined—according to which research? based on which assumptions?—effectively maintains the status quo in policy decision making that led to the hazard in the first place. One must consider how safe "BPA-free" is. That goes not just for consumers but for retailers and manufacturers as well, who most likely want to choose a better substance to substitute for a potentially hazardous one.

The answer to the question "Is BPA safe?," therefore, is not as simple as yes or no, nor is the solution absolute restriction without consideration of what will replace it. Yet as the stakes in the debate rose the issue became increasingly polarized: safe or not safe, ban or no ban. To expand the issue and build on the debate, advocates sought to point to BPA as an example of regulatory failure. As with past "policy tragedies"—the problems highlighted in Upton Sinclair's *The Jungle*, which preceded the Pure Food and Drug Act of 1906, and the thalidomide crisis, which provided the political momentum necessary to pass major drug reforms in 1962—BPA was part of a larger political call for regulatory reform. In the "Baby's Toxic Bottle" report, advocates called not only for bans and market shifts but also for broad-based chemical reforms. As consumer anxiety about BPA in baby bottles, lead in toys, and phthalates in rubber ducks rose, public confidence in the federal government declined.[60]

CHEMICAL POLICY REFORM

By the beginning of the Obama administration, a number of states, including Washington, Maine, and, most notably, California, had passed legislation that aimed to prioritize, evaluate, and regulate hundreds of chemicals in commerce. In 2008, Arnold Schwarzenegger, governor of California from 2003 to 2010, signed into law two bills (AB 1879 and SB 509) that created the Green Chemistry Initiative. While "green chemistry" includes twelve principles of design first articulated by chemists John Warner and Paul Anastas, the California law lays out a regulatory policy for existing and new chemicals.[61] Under the new laws, the California Department of Toxic Substances Control has the authority to establish a process to identify chemicals of concern and a process to evaluate those chemicals in order to limit or reduce exposure. In addition, a clearinghouse of information on chemicals is to be established and maintained.[62]

This comprehensive legislation grew out of consumer and advocacy pressure as well as subsequent market uncertainty. Individual chemical bans served as test cases for passage of larger reforms, and collectively such state insurgencies threatened to introduce a patchwork of chemical regulations across the United States that could raise the costs of doing business. Moreover, the legal confusion and financial burden of establishing different regulations in different states generated political and economic pressure for federal reform. The federal statute targeted for reform was the TSCA.

Signed into law by President Ford just before he left office in 1976, the TSCA provides the EPA with the authority to obtain information on industrial chemicals, including health and safety, production, and use data, and to control or restrict any chemical that poses an "unreasonable risk" to human health or the environment.[63] Unlike the Federal Food, Drug and Cosmetics Act of 1958, the TSCA does not require premarketing testing, and it effectively grandfathered in the vast majority of chemicals on the market (i.e., "existing chemicals") with little to no safety data. In the three-plus decades since its passage, EPA has issued regulations on only five substances: PCBs, chlorofluorocarbons, dioxins, asbestos, and hexavalent chromium. And these rulings did not come easily, partly because of the difficulty EPA experienced in obtaining data on existing chemicals and issuing regulatory standards.[64]

The leading example of the statutory flaw in the TSCA is that of the asbestos ruling. For ten years the EPA worked on asbestos risk assessment and regulatory rule making as a test of the major provisions in the law. The agency issued a regulation that would have restricted asbestos in almost all uses, and asbestos producers, already under profound liability pressure, immediately challenged the decision in court. In the 1991 case *Corrosion Proof Fittings v. EPA*, the Fifth Circuit Court of Appeals overturned the EPA's asbestos standard, ruling that the agency had failed to meet the high burden of proving with "substantial evidence" that an "unreasonable" risk existed. Further, the court found that the agency had not sufficiently demonstrated that the proposed regulation presented the "least burdensome" approach, as directed under the TSCA.[65] Meeting the "least burdensome" standard would have required much more extensive evaluation of the risks and benefits of all substitutes for asbestos. The court's decision struck a major blow to the EPA's authority and reputation. The agency chose not to appeal the decision, and the morale of EPA staff involved in the long assessment plummeted.[66]

The implication of the failed asbestos regulation was that the agency's authority over existing chemicals was profoundly limited. Since the 1990s, legal experts, environmental advocates, and the U.S. Government Accountability Office (GAO) have issued a number of reports on the central flaws of the TSCA. Among its major failures is the difficulty the EPA faces in obtaining data necessary to evaluate existing chemicals. To obtain data, the EPA must first determine that a chemical "may present an unreasonable risk" to human health or the environment, or that human exposure or environmental release is significant. This has created what some have called a Catch-22 situation: the agency can't make a finding of a potential risk or significant exposure without data, but it can't get data without evidence that an "unreasonable risk" may exist.[67] "Typically with little or no data, EPA is required to make formal findings about the potential toxicity, the adequacy of other federal laws, and alternative options before it can demand that specific tests be conducted."[68] Moreover, the broad use of what is called "confidential business information" has arguably further weakened the implementation of the law by limiting the agency's ability to share publicly information on chemical use, production, and risks.[69] The severity of the flaws in the TSCA was highlighted in 2009 when the GAO identified the statute as an area of "high risk" for waste, fraud, abuse, and mismanagement and a national priority for reform in the High-Risk Series, a biennial report to Congress that lists critical priorities for legislative action.[70]

The chemical industry trade associations, by contrast, have long maintained that the TSCA is a highly effective statute.[71] After all, it has generated few data burdens, and with such a high burden of proving that an existing chemical presents a significant risk, almost all chemicals have remained on the market. The industry's position on the TSCA began to subtly change by 2008, given new demands for chemical data from a number of U.S. states, as well as Europe. For the first time ever, the ACC publicly outlined a number of "principles for modernizing TSCA."[72] In

Europe, implementation of a major new chemicals management program in 2007—Registration, Evaluation, Authorisation, and Restriction of Chemicals, or REACH—promised to generate considerable demands for data from chemical producers in order to maintain or obtain market access.

REACH was the product of eight years of negotiations among chemical trade associations, the U.S. government, European governments, and European NGOs. The regulatory situation for chemicals in Europe prior to REACH was similar to that in the United States under the TSCA. Most notably, as in the United States, the vast majority of chemicals on the European market had been grandfathered in, with little to no information on their safety or toxicity. European NGOs formed a broad coalition that included Friends of the Earth, Greenpeace, the World Wildlife Fund, the European Environmental Bureau, and Women in Europe for a Common Future to advocate strong reforms that would demand data for all chemicals in production, place the burden on industry to generate that data, and incorporate a precautionary approach to chemicals management.[73] The EU is the largest global chemical market, so any reform there could have significant impact for most U.S. producers seeking to import into the region. During the George W. Bush administration, the U.S. trade representatives, Commerce Department, State Department, and EPA coordinated efforts to block the European reform and, when that failed, sought to significantly water down the data requirements in REACH.[74]

Eventually passed in 2006, REACH established a complex management program for all industrial chemicals produced in or entering Europe. REACH requires data on safety for both new and existing chemicals on the market and places the burden of proving safety on the industry. Chemicals must be "registered" (R), "evaluated" (E), and "authorised" (A) for use by the European Chemicals Agency (ECA), the agency established under REACH. Chemicals not given authorization can be restricted from the market. The law created what is commonly referred to as the "no data, no market" rule. REACH created unprecedented rules for data generation by industry and information sharing in the marketplace (e.g., use information is shared up and down the production chain).[75]

Given that there are tens of thousands of chemicals in production, the European Commission recognized that not every compound would be reviewed by the ECA. To expedite the review and regulatory standard-setting process for chemicals, REACH gives the ECA the authority to establish a list of priority chemicals. This "candidate list" includes what are referred to as "substances of very high concern," defined as chemicals with carcinogenic, mutagenic, or reproductive toxicity properties (CMRs) and/or persistent, bioaccumulative, and toxic compounds (PBTs), as well as individual chemicals considered to be of an "equivalent level of concern," which REACH specifies to include endocrine disruptors. On the basis of reviews of the data, those "substances of very high concern" may then be placed on an authorization list. If a chemical is on that list, a company must obtain prior approval for production. Approval can be obtained by demonstrating one of two conditions: first, that the risk of the substance is adequately controlled; or, second, that the chemical's socioeconomic benefit outweighs its risk.[76]

These two conditions reflect the fundamental pillars of international chemicals policy, which are supported by the chemical industry. First, because control is a form of exposure reduction, even a chemical placed on the authorization list because of significant hazardous properties may continue to be produced if assurances are given that exposure is safe and can be controlled at low doses. This provision provides an opportunity to maintain the market for a chemical of concern. Second, in determining market access, one must balance the economic benefits of production against the chemical's risk.

While the concept of "no data, no market" seeks to fill the tremendous gap in knowledge about chemicals in commerce, defining safety remains at the heart of the debate over implementing REACH. While obtaining more data on chemicals promises to provide, at a minimum, some information on toxicity, determining which studies are conducted (e.g., which doses will be used, which methods, and based on which assumptions about biological activity), the quality of that data and the interpretation of the findings remain points of debate in arguments over whether a chemical should be allowed on (or allowed to remain on) the market. Further, REACH's data demands have provoked considerable debate on the economic feasibility of testing and the tremendous costs in terms of animal lives sacrificed.[77] (Animal rights organizations were strongly opposed to REACH data requirements and have sought to limit testing programs, such as the Endocrine Disruptor Screening Program, in the United States.)

The first years of the Obama administration, when the Democratic Party controlled both houses of Congress, were a window of opportunity in which to pass major legislation on a wide range of issues, from climate change, health care, and immigration reform to regulation of toxic chemicals. Senator Frank Lautenberg (D-NJ) and Representatives Waxman and

Bobby Rush (D-IL), with support from a new coalition of environmental, consumer, and health advocacy organizations known as the Safer Chemicals, Healthy Families campaign, introduced the Safe Chemicals Act to substantially reform the TSCA. The new legislation aimed to improve the EPA's ability to obtain data for chemicals on the market (with the burden on industry to prove safety, as required under REACH), to expand public disclosure of data, and to speed up risk management of those chemicals for which extensive risk assessments exist, including chemicals classified as persistent, toxic, and bioaccumulative. The bill served as a starting point. Lautenberg introduced another, slightly altered bill, the Safe Chemicals Act of 2011, with many more Democratic cosponsors, in an effort to build political momentum to move the legislation through the Senate.

In the House, Representatives Jan Schakowsky (D-IL) and Edward Markey (D-MA) introduced the Safe Cosmetics Act, legislation that would overhaul the cosmetics law, never amended since its original passage in 1938. The introduction of the cosmetics bill, which has yet to come up for serious debate or negotiation, draws its political support from an environmental health advocacy campaign for safer cosmetics: the Campaign for Safe Cosmetics, organized by a coalition of NGOs including the Environmental Working Group and the Breast Cancer Fund. In 2005, pressure from this campaign led to the successful passage of a California law that requires the listing of, rather than the restriction of, carcinogens, mutagens, and reproductive toxins in cosmetic and personal care products sold, manufactured, distributed, or packed in the state. Once again, action in Europe provided significant market pressure. In 2003, the EU passed the Cosmetics Directive, which restricts chemical carcinogens, mutagens, and reproductive toxicants from personal care products: soaps, shampoos, gels, cosmetics, lotions, and so on.[78]

Politically divisive debates that erupted over health care reform, marked by hysteria over "death panels," the failure of a bipartisan cap and trade climate bill, the rise of the Tea Party movement, and subsequent congressional intransigence in the debate over the federal budget, effectively eroded away hope of legislative activity on nearly any issue. Yet although climate change legislation became a politically moribund issue, toxic reform, which had always been a secondary legislative priority for the environmental community, quietly and slowly continued to make progress through dialogue with chemical companies and industry trade associations. With a national election looming as this book goes to press, however, the prospect for reform remains uncertain, as no Republicans have publicly stepped out to co-sponsor the legislation.

INTERPRETING SAFETY, REFORMING REGULATION

Over the past several years, state bans on BPA and calls for broader reforms of the TSCA have signaled a growing popular perception of the federal government's failures to protect public health from hazardous chemicals. Responsibility for that failure, in the case of TSCA reform, is placed on the weakness of the statute itself, and thus the political response demanded is action by Congress. In the more narrowly defined case of state bans of BPA, however, blame for regulatory failure falls on the FDA, and therefore the action demanded must come from the agency itself. Both narratives of failure point to the considerable challenge regulatory agencies have experienced in attempting to restrict even extensively studied chemicals such as asbestos and BPA. Proposed TSCA reform would attempt to resolve some of this challenge by expediting the risk-management process for heavily studied and assessment chemicals. What, then, is the proposed reform for the FDA?

In the case of BPA, despite the volumes of papers published on the chemical, the FDA continues to uphold its safety. Demands for FDA action to reevaluate BPA presume that the agency itself views its decision as a failure. But if there is no significant change in which information is used to inform decision making, or in the testing protocols used in standardized testing systems, as recommended by the NTP over a decade ago, then the agency's perceived failure will persist and it will find itself under continued scrutiny by and pressure from advocacy groups.

The FDA began to come under congressional and advocacy scrutiny at the end of George W. Bush's term in office. As part of the congressional investigation into BPA, Representatives Dingell and Stupak sent a letter in 2008 to the commissioner of the FDA, Andrew von Eschenbach, inquiring about the agency's position that there existed "no reason at this time to ban or otherwise restrict uses now authorized."[79] Specifically, Dingell and Stupak's letter asked von Eschenbach on which studies the agency based its claim that there was no "safety concern at the current exposure level."[80] In response to this pressure from Congress, in the final months of the Bush administration the FDA issued a draft reassessment of BPA's safety that upheld its existing safety standard at 50 µg/kg/day.[81]

The agency based the standard on evidence from the two multigenerational studies funded by the American Plastics Council and the SPI.[82] For FDA staff, many of whom stayed on to serve in the Obama administration, these studies allowed for a straightforward and familiar approach to determining a safety standard. The tests evaluated well-known toxicological endpoints, used a range of doses, exposed the animals orally, tested a large number of animals, and followed GLP. As for the research on low-dose effects of BPA that raised "some concern" for the NTP, the FDA initially found these studies to be inadequate or of limited utility in evaluating safety. The agency pointed to the smaller number of animals, the limited number of endpoints, the nonoral dosing, and the narrow windows of exposure as reasons to disregard most of the academic research.[83] The agency needed large regulatory studies that used large numbers of animals, a wide range of dosing to determine a dose-response relationship, and conventional endpoints well recognized as toxicologically adverse.

The agency's interpretation of the academic research as entirely unfit for regulatory decision making and its continued reliance on the two large Tyl studies placed it in general agreement with industry trade associations. The FDA also drew on a number of industry-sponsored reviews of the literature, including a review of developmental neurotoxicity by Oak Ridge National Laboratory. The report included a literature review submitted by Steve Hentges of the American Plastics Council, conducted by the firm Exponent and paid for by the ACC.[84] In 2009, reporters at the *Milwaukee Journal Sentinel* requested copies, under the Freedom of Information Act, of all correspondence during the period in which the BPA report was prepared by the agency and detailed frequent exchanges and consultation between the FDA and the ACC.[85]

The agency's reliance on the chemical trade association for expert advice and scientific interpretation was not an aberration of the Bush administration. Rather, the relationship demonstrates the long-standing success of the chemical and plastics industries' efforts to become leading experts in the science of toxicology, to maintain toxicology as the scientific discipline for evaluating chemical safety within the regulatory context, and to foster a working relationship with the regulatory agency, regardless of which political party takes control of the executive branch.[86] The implication is not simply that the agency has been captured by the regulated industry but that over time a mutually shared process to assess and interpret research has emerged that has helped to minimize conflict and ease the regulatory process for both sides. Moreover, a community of people working on chemicals in foods regulations that includes industry and FDA officials has developed over time as a result of shared work experience. For example, after working on the contentious BPA reassessment, one FDA staffer did what many before her had done: she took a position—most likely a more lucrative one—at the law firm Keller and Heckman.[87]

The accepted production and flow of information for regulatory decision making were outlined by John Rost, chair of the North American Metal Packaging Alliance, in a presentation in 2010. He stated that validated regulatory tests, funded by industry and following GLP, should directly inform regulatory decisions. More exploratory, peer-reviewed research conducted by academics, he noted, might inform these regulatory safety tests but should not directly influence the decision-making process.[88] This prescribed production and use of information, which disallows experimental research from directly informing decision making, effectively conflates sound science with regulatory science and, by proxy, with research paid for by industry. Managing the definition of and propagating the production of sound science give the chemical industry effective control over which evidence will be used to determine safety and measure risks.

This accepted process explains why regulatory agencies around the world have come to the same conclusions as the FDA about BPA's safety. A review by independent Swedish researchers of risk assessments conducted on BPA in Europe, the United States, Canada, and Japan found that all of the assessments relied heavily on the two multigenerational Tyl studies, funded by industry and conducted at a private laboratory in Research Triangle Park, North Carolina.[89] These two studies figured prominently in the Annapolis Center's weight-of-the-evidence reviews conducted by the Harvard Center for Risk Analysis and Gradient Corporation. When the European Food Safety Authority (EFSA) evaluated the safety of BPA in 2002, the agency raised the standard to 50 ug/kg (the FDA standard) on the basis of these two studies; it reevaluated BPA safety in 2006 and upheld the new safety standard.[90] This harmonious process and relationship were disrupted by emerging research into and public outcry over BPA.

The first strong criticism of the FDA's review of BPA's safety came in 2008 from a Scientific Board Subcommittee of the FDA. Composed of outside experts, the subcommittee was tasked with evaluating the agency's report on BPA, a process that required public input. At the first public meeting, the subcommittee heard presentations from John Bucher, head of the NTP, who reviewed the conclusions of the BPA monograph, as well as from Fred vom Saal, who discussed the findings of the Chapel Hill Consensus Statement on BPA. In its final report to the FDA, the subcommittee called for major revisions to the report. It disagreed with the agency's decision to exclude all non-GLP studies from its assessment and raised serious concerns about the agency's exposure assessment, which one member unequivocally stated was seriously flawed.[91]

The flawed assessment went back to the agency for revision just as the Obama administration came into office and issued a promise to restore scientific integrity throughout the government.[92] A former New York City health commissioner, Margaret Hamburg, became commissioner of the FDA in 2009 and immediately called for reforms to regulatory science, including the updating of regulatory tools and methods.[93] As part of this agency-wide initiative, the FDA contracted with Lynn Goldman, former assistant administrator at the Office of Prevention, Pesticides and Toxic Substances during the Clinton administration, to advise the agency on its BPA review.

The agency's initial response under its new political leadership was twofold. First, it announced a five-year research plan for BPA designed to clarify several points of scientific controversy and to take a step toward updating testing methods and protocols. This included studies of neurological effects, about which the NTP reported "some concern," and pharmacokinetic studies in order to better understand how BPA is metabolized in the body.[94] In addition, the NIEHS announced investments of $30 million over two years for BPA research that would examine some new endpoints of concern in larger studies designed for safety assessment. For instance, the studies would measure new endpoints across a wide range of environmentally relevant exposures, orally exposing the animals and using large numbers of animals.[95] The intent, in part, was to use credible evidence from academic research to inform the design of new toxicity testing so as to make it more relevant and up to date. Funds were made available through the American Recovery and Reinvestment Act.[96] (The published results of most of these studies were still pending when this book went to press.)

The second agency response was its announcement that it would adopt the NTP's conclusion of "some concern" for effects on the prostate, brain, and behavior in fetuses, infants, and children at current levels in humans.[97] Though moderate and measured, the shift in tone was significant. The agency's reliance on industry-sponsored studies as the sole relevant and reliable data to determine safety appears to have been overturned, at least temporarily. The existing safety standard for BPA, however, remains.

On the heels of the FDA's announcement, the EPA issued a "chemical action plan" for BPA.[98] Early in her tenure in the Obama administration, EPA administrator Lisa Jackson declared that the agency was back in business when it came to enforcing chemical and environmental laws.[99] The "chemicals of concern" list of eight substances, including BPA, signaled a more aggressive effort by the agency to enforce the TSCA.[100] The EPA identified BPA "as a substance that may present an unreasonable risk of injury to the environment on the basis of its potential for long-term adverse effects on growth, reproduction and development in aquatic species at concentrations similar to those found in the environment."[101] Observing jurisdictional boundaries, the EPA left the determination of safety as it directly related to humans to the authority of the FDA, focusing instead on plans to gather and evaluate environmental effects data as well as to assess chemical alternatives to reduce BPA use and emissions. This included an "alternatives assessment" of thermal, or carbonless, papers—used, for example, in receipts—that examined the different hazard properties of BPA-based papers compared to available alternatives. In recent years, exposure to BPA from thermal paper has drawn considerable attention because the concentrations of the compound may be extremely high (10 milligrams). However, because the exposure route is through the skin, human exposure has been estimated to be far lower than through the diet.[102]

For Senator Feinstein (D-CA) and some environmental health NGOs, these moderate regulatory steps—more research and evaluation—were inadequate and slow responses to a problem of immediate public concern. By 2010, passage of state laws to limit or restrict BPA had generated considerable momentum for a showdown in the Senate. Senator Feinstein sought to include a BPA ban as an amendment to the Food Safety Modernization Act, a massive bill designed to overhaul food production and processing regulations in response to a series of deadly foodborne illnesses transmitted by tainted products such as peanut butter and spinach.[103] Food and grocery trade associations balked at the amendment, arguing that it threatened the future of the bill. In a public letter to Feinstein, Representative John Dingell, who authored and helped pass the House bill, asked the senator to drop the amendment.[104] Feinstein conceded, and the bill was signed into law in 2011 without the BPA amendment.

The subtle yet significant shift in the position of the Obama administration's FDA and EPA—a break from their absolute confirmation of BPA's safety—and consistent legislative pressure had a ripple effect across the Atlantic in Europe. In 2010, Denmark and France passed restrictions on BPA in children's products, which, like Canada's, explicitly articulated the decision as an implementation of the precautionary principle. As Henrik Hoegh, the minister of food in Denmark, noted: "In my opinion these uncertainties [neurological effects in rats] must benefit the consumers, so we will utilize the precautionary principle to introduce a national [temporary] ban on bisphenol A in materials in contact with food for children aged 0–3 years."[105] The French National Assembly went a step further by passing a full ban on BPA in food packaging.[106] The actions by Denmark and France triggered further review by the EFSA. The FDA and EFSA have been in similar agreement about the safety of BPA. However, in reviewing the EFSA report, the European Commission's health and consumer policy commissioner noted a number of uncertainties about BPA's effects on tumor promotion, development, and immune response, "which will have to be taken into account in possible action to be taken by the Commission."[107] It should be noted that the report also included a minority opinion, which highlighted considerable uncertainties in the validity of the current "no observable adverse effect level" (NOAEL) due to new research on prenatal and neonatal effects published since 2006. "The effects of concern include, in particular, brain receptor programming, immune modulation and enhancement of susceptibility to breast cancers." Since none of these studies could be used to establish a new safety standard, the minority opinion suggested that the current standard be accepted as "temporary." The minority opinion targeted the panel's invalidation of several studies to distinguish that such studies were not invalid but were instead unsuitable for setting a new safety standard. The minority report's additional recommendations included a call for more robust research, particularly in areas not addressed in testing guidelines, such as metabolic and immune effects.[108] Once again, the gap between regulatory safety tests and smaller, more focused academic research could explain the difference in how a reviewer interprets the potential for risk.

Given the persistent uncertainty about the human health risks of BPA and the conflicting policy responses it was provoking across the globe, including limited bans in China, the World Health Organization (WHO), together with the Food and Agriculture Organization of the United Nations, convened an expert, invite-only meeting in 2010. The meeting brought together experts from around the world and included controversial figures such as vom Saal. As in the NTP monograph, the WHO report evaluated the weight of all the existing evidence (e.g., exposure assessment, biomonitoring, epidemiological studies, reproductive toxicity, metabolic effects, mechanisms of action) and came to similar conclusions. Like the NTP/CERHR report, the WHO report reaffirmed the certainty that BPA does not cause adverse effects on "conventional" endpoints at low doses. However, the report acknowledged, as did the NTP/CERHR report, some evidence of effects on new endpoints at doses relevant to human exposure levels, including behavioral and possible metabolic effects. Importantly, the report noted that the prevailing reliance on the estrogenic mechanism of action is probably not relevant for evaluating such effects. Yet many of the effects of BPA have been evaluated according to its interaction with classic estrogen receptors, which could lead to inaccurate conclusions. While the compound is estrogenic, there is good evidence that its biochemical activities are very complex and include other receptor systems (e.g., androgens, thyroid, glucocorticoid, and many other signaling systems). This means that evaluating effects on the basis of assumptions of consistency with estrogenic responses may be inappropriate for certain systems and/or endpoints.[109]

Such mechanistic complexity, combined with the questionable reliability of many research studies and the unresolved controversy about the significance of some effects measured in the studies, the report concluded, continued to plague the regulatory process.[110] While such complexity might help to inform new directions in research and provide some explanation for the uncertainty of effects, it provided no resolution to the policy predicament of whether to maintain the default position on BPA's safety. In other words, the gap between academic research and standardized regulatory tests continued to make translation of the risks to public health exceedingly challenging for European and U.S. regulatory bodies.

BPA bans and the concomitant decline in consumer confidence in product safety drove retailers and producers away from this chemical, even while federal agencies around the world declared the compound safe. With so many BPA bans on baby bottles in the United States and Europe, in early 2012 the ACC actually requested that the FDA revoke the acceptable use of BPA-based polycarbonates for baby bottles and sippy cups. The reason given was not the toxicity of BPA but the ACC's contention that the market for BPA-based polycarbonates in these products had been abandoned.[111]

While it is tempting to declare the regulatory process hopeless, the market made only a subtle shift after sustained public interest and considerable government investment in research. This is not a tenable or sustainable approach to dealing with the hundreds of compounds that may pose significant public health risks. Further, some researchers now express concern about the compound BP-S, which appears to have replaced BPA in thermal paper.[112] Without changes in regulatory testing methods and requirements for such data, one chemical of concern may have simply been replaced with another. Producers and consumers need a flexible, independent, and transparent regulatory agency to set rules and provide guidance on risks, using the best available research. Proposed legislative changes under the TSCA hold some promise for increased public disclosure of data on chemical use, toxicity data, and production. Although having data is a necessary first step, it is not sufficient to move markets or protect public health. Which information is produced and how that information is interpreted and used in decision making will determine which chemicals remain in production and, ultimately, in the human body.

A MODEST PROPOSAL AND CONCLUSION

The question of how to determine the safety of BPA will not be solved with more data alone. Nor will it suffice to shift the burden of proof to industry. In the case of food contaminants such as BPA, the burden has been on industry to demonstrate safety since the late 1950s. Promoting BPA as a poster child for reform, therefore, has generated cognitive dissonance in the chemical debate. The advocacy campaigns demand safer chemicals, but by working through the legislative process to ban BPA and bypassing the FDA, the regulatory decision making process to define safety is left largely unchanged.

It is through the process of interpretation that the meaning of safety is defined. Which research is conducted and how studies are evaluated, by whom, and with what degree of transparency all contribute to a picture of a chemical's risk to health. One study, no matter how large, cannot paint the complex landscape of what is and is not understood about risk. When value judgments involved in interpreting data are not made visible and highly transparent, disagreements over personal or ideological bias can become conflated with scientific debate about whose science is more valid or sound.

In the case of BPA, concerns about private-industry influence on scientific interpretation and regulatory decision making erupted over the FDA's persistent reliance on industry studies to uphold the safety standard. When the FDA argued that studies that followed GLP standards were sounder science for regulatory decision making, the agency made a number of problematic assumptions. The GLP standards, adopted by private contract firms in the 1970s, provide research guidelines to enhance the validity of data and the reliability of methods. While GLP unequivocally provides necessary standards for production of data, the FDA conflated consistency in the production of data (i.e., reporting validity) with the assumption that the best available tools were being used and the appropriate and relevant questions were being posed (i.e., issues that relate to the internal validity and external validity of a study). But what good is it to have reliable data about the quality of a patient's heart if the problem might be with the liver? Moreover, by assuming that two studies could tell a whole picture of a chemical's safety, the agency traced the cognitive pattern outlined by industry.

But industry needs a stable and consistent process to define safety. Maintaining the stability of the questions asked about a chemical's toxicity is an important way to ensure such consistency. While it may be incumbent upon the private sector to conduct or sponsor research, it is not in the public's interest that industry should determine which questions are asked or how research is interpreted. When industry manages the interpretation of science by determining which research is valid and which studies are relevant for determining safety, it effectively captures the critical responsibilities of the regulatory agency. It is the agency's responsibility to adjudicate among different assumptions, values, and perceptions and to interpret the weight of the evidence. Absent from the agency's decision making process on BPA was a clear, transparent set of criteria for how it evaluated the existing body of research.

How, then, might the process of interpreting research better account for inequitable values and financial conflict of interests? One possibility is to advocate a continuous need for vigilance, of the sort practiced by consumer advocates in the late 1960s and 1970s and environmental health advocates of today. Public interest groups have effectively used multiple channels of political influence to make conflicting interpretations of safety and risk visible and to hold regulators accountable. The tools and methods deployed have included successful use of the media and the support and lobbying of local, state, and national lawmakers; both entities are responsive to their constituencies and capable of intervening in science policy and regulatory decisions. These strategies, used by many social movements, illustrate an advantage of a democratic society: alternative pathways to check economic and political power. Yet there are limits to the effectiveness of public participation in determining the safety of chemicals. Neither journalists nor members of Congress are qualified to arbitrate scientific disputes. That is the role of scientists and regulatory agencies.

The question, then, becomes whether the values of a democratic society, including processes to check power and ensure transparency, can be integrated into mechanisms to evaluate conflicting or diverse bodies of scientific evidence. Here there may be lessons from the history of health policy and the emergence of systematic reviews of evidence.

The gold standard for evaluating a drug's clinical effectiveness is the randomized controlled trial. For decades in the United States, however, most randomized controlled trials were funded by the pharmaceutical industry as part of the process for drug approval.[113] As Dan Fox details in his recent book on the subject, growing concerns about financial conflicts of interest in drug research and demands for better understanding of effective health care, given rising health care costs, generated interest in the advancement of methods for systematic review of clinical research among researchers and policy makers. The objective of a systematic review is to provide a transparent, replicable method to collect and evaluate existing evidence to answer a predetermined research question.[114] The process by which studies are selected and evaluated is designed to be transparent and rigorous, with attention to the evaluation of different study biases, including conflicts of interest.

In the late 1980s, for the first time, systematic review methods were applied to an entire area of health care: obstetrics. Iain Chalmers at Oxford University published a massive systematic review of the evidence in practices and interventions in pregnancy and childbirth, *Effective Care in Pregnancy and Childbirth*. Working with partners across the world, Chalmers led the development of the Cochrane Collaboration in the early 1990s. Organized as an international nonprofit organization, the Cochrane Collaboration formed to conduct, maintain, and disseminate systematic reviews of research in medicine and health care through the creation of an online journal, the *Cochrane Library*.[115]

The organization's namesake and one of its founders, Archie Cochrane, had previous experience conducting independent randomized controlled trials and developed methods to evaluate and manage bias in reviewing evidence. Independence and transparency are key characteristics of systematic reviews conducted as part of the Cochrane Collaboration. Notably, the Cochrane Collaboration prohibits authors of the systematic reviews it publishes from receiving funding from pharmaceutical companies or other commercial entities. Funding for the Collaboration comes from governments, hospitals, universities, foundations, individual donors, and proceeds from the *Cochrane Library*.[116]

Over the past twenty years, a number of guidelines for systematic review have emerged along with the establishment of dedicated journals.[117] The process is clearly evolving and iterative. As some early participants in the Cochrane Collaboration noted in the mid-1990s, developing credible reviews of all sectors of the health care system would require constant modification, monitoring of data quality and integrity, and mediation of conflicts and disagreements.[118] More recently developed methods include the assessment and ranking of recommendations or decisions based on systematic reviews. Such methods aim to transparently outline various decisions based on what is known and not known.[119]

Drawing from the experience of and methodologies developed by the Cochrane Collaboration, a consortium of researchers, led by Tracey Woodruff at the University of California, San Francisco, are developing a systematic review methodology to evaluate environmental health research.[120] Together with partners from state and federal agencies, including the EPA, the NIEHS, and veterans of the Cochrane Collaboration, Woodruff and her colleagues launched the Navigation Guide in 2011, a systematic review and decision-making tool for environmental health research. Like other systematic review methods, notably the Campbell Collaboration, used to evaluate research in other fields, such as education, criminal justice, and social welfare, the Navigation Guide project will require modification and adaptation to the very diverse streams of evidence used in environmental health. Perhaps the most important factor and challenge is that, unlike in the clinical setting, where human observational studies and randomized controlled trials are common, most evidence in environmental health comes from animal studies that frequently use varying methods and study designs. As this book was going to press, Woodruff and her colleagues were working on a number of case studies to test the methodology as well as developing a network or collaboration for advancing the utility and use of the tool.[121] Systematic review methods and several case studies were also being advanced at the NIEHS's Center for the Evaluation of Risks to Human Reproduction, appropriately renamed the Office of Health Assessment and Translation.

While not a panacea to the crisis of defining chemical safety, systematic reviews of environmental health research address several critical problems. First, systematic review makes the process of collecting and evaluating data highly transparent. Second, it allows for rigorous evaluation of different biases that can affect a given study (i.e., systemic error in a study). This second aspect does not apply a single value to a given study—for example, useful, not useful, poor, good, valid, invalid, and so forth—which collapses many different evaluations into one score. For example, a small study with few subjects tends to have greater variation and thus less precision than a study that uses a greater number of subjects. This imprecision, however, is not the same as a systemic bias. If a study used an insensitive or inappropriate test to evaluate a research question, regardless of the sample size, the study would have a systemic error. Even with large numbers of animals, the study would arrive at the wrong answer, albeit with greater precision. Therefore, if a study's utility was heavily weighted on sample size alone, such a large study could be considered useful despite having a systemic error.[122]

Prying open the evaluative process further enhances transparency and allows for many different researchers to explore the data. This transparency and greater independence of evaluation reflect important principles of democracy and the pursuit of truth. A systematic review process could provide reliable and trusted knowledge about what is known and not known about a chemical's risk from which policy makers could draw when adjudicating scientific conflicts. Systematic reviews also reveal serious gaps in knowledge and research. Revealing what isn't known is an important part of decision making notably absent from the current process of defining chemical safety, in which the absence of evidence is often interpreted as the absence of risk.

Other models of independent review within the environmental health arena do exist. In 1980, the Health Effects Institute was created to evaluate research on the health effects of air pollution, a highly contentious area of the debate over regulations for clean air standards. The institute receives half of its funding from the EPA and half from private industry. An independent board of directors oversees a research committee and a separate evaluation committee.[123] By managing the problem of financial conflicts of interest, the Health Effects Institute and the Cochrane Collaboration explicitly recognize that those with commercial and pecuniary interest in protecting a product or market can influence the production and interpretation of data. Rather than pretend this is not the case, it is more realistic and pragmatic to acknowledge and manage it.

Today, battles about the safety of BPA and other controversial chemical compounds, including phthalates and flame retardants, continue to destabilize markets, challenge the authority and power of regulatory agencies, and confuse consumers. Bans on chemicals are a reactionary approach to public concern. Politically, a ban gives advocates a target. The chemical becomes an antagonist in a bigger story about the failure of the regulatory system. But banning BPA alone won't fix a larger systemic problem detailed in this book: how to define safety in the face of advancing and shifting scientific knowledge and competing efforts to interpret that science.

The complex question "Is it safe?" demands more than a one-word answer. Developing independent methods and means to evaluate what we know and don't know about chemicals and building communities of researchers to translate the body of evidence can offer an open and collaborative process to navigate safety. This is not a solution to scientific uncertainty or the inevitability (and necessity) of scientific debate. The goal is to provide individuals, policy makers, and clinicians with trusted sources of information to assess different recommendations for reducing risk and improving public health. Ultimately, the hope is to protect health and strengthen democracy in science.

Epilogue

In a 2009 episode of *The Simpsons*, BPA became fodder for prime-time parody. Banished from making snacks for her mothers' playgroup—a group of educated and internationally diverse women that might represent today's self-described "eco-moms"—Marge Simpson decides to convert the Simpson family to a healthy, socially conscious diet. The first step is a trip to a grocery store that looks conspicuously like Whole Foods, where the Simpsons rack up a ludicrously high bill familiar to any shopper at high-end natural food stores. Next, Marge bakes what she conceives to be the ultimate healthy, socially conscious, safe snack food: "homemade, organic, nongluten, fair-trade zucchini cupcakes." She presents it to the other mothers, and when they ask her what amount of unsalted butter she used to grease the pans, she proudly announces none, because she used a nonstick pan. All of the mothers immediately spit out the cupcakes with revulsion and chastise Marge. Didn't she know that nonstick pans were made with PFOA! "There is only one thing more dangerous than PFOAs, Marge," one mother exclaims. "Plastics made with BPAs. Never, ever let your child near any product with the number 7." Dramatic music plays as a number 7 appears at the bottom of a sippy cup used by one of the children. "Good Lord," exclaims another mother, "they've been sucking 7. Ahhhhh!" The cup is knocked from the child's mouth and the entire playgroup flees the Simpsons' house in hysteria. Running out into the street, they flag down an ambulance, toss an injured man from the back, and jump in. The ambulance tears off down the street.[1]

Public perceptions of the risks of BPA had been reduced to the cause of hysterical, overprotective, and overeducated mothers. Was this the inevitable counterreaction to a persistent environmental campaign that had indeed targeted mothers? Despite recognition among most environmental advocates that humans are exposed not just to BPA but to a complex mixture of industrial chemicals, the political symbolism of BPA proved to be an irresistible target. Mothers' choices move and make markets. When you tell parents, state legislators, and members of Congress that their children are born polluted, they listen and take action, whether through market decisions or legislative bans. BPA served as a seemingly perfect example of how chemicals get into the body and alter it in ways we'd failed to consider.

But the question of whether BPA is safe does not address the larger issue of this book: What is the meaning of safety? The rapid proliferation of research revealing the presence of industrial chemicals in our bodies, their complex interactions with genetic expression, and their influence on biological development and function will continue to fuel debate and confusion about risks. We have become what we make, but we lack the means to determine and effectively manage the long-term risks of this reality. In *World at Risk*, Ulrich Beck describes the challenges of environmental deterioration and argues that the "foundations of the established risk logic tied to the nation-state are being undermined."[2] In other words, the systems developed to protect against the possibilities of harm and calamity no longer fit the risks of today. In the case of chemical hazards, the logic of risk reduction and management developed over the past half century falls short in mitigating contemporary chemical risks. Specifically, the assumption that safety exists at low doses is falling apart as research explores more fully the industrial chemicals in our bodies and their biological impacts. The debate about BPA's safety demonstrates how scientific research is crashing up against our models for organizing and distributing risk in society.

A number of efforts now under way suggest that a transformation in the methods and means used to identify, manage, and mitigate chemical risks has begun. Among these changes are recent recommendations to modernize risk-assessment methods. In 2007, the National Research Council issued an updated report on risk assessment, *Science and Decisions*, that addressed a number of factors contributing to a perceived crisis in the methodology. This crisis was a result of a number of converging factors, including increased scientific complexities, conflicting interpretations by diverse stakeholders, and political challenges to the method's credibility as an effective decision-making tool due in part to the long time frame and disputes that characterize the process. In this respect, the report addressed both the technical and nontechnical aspects of risk assessment.[3]

As Garry Brewer notes in a review of the report, *Science and Decisions* directly confronts the nontechnical aspects of risk assessment that contribute to the problem of decades-long assessment processes, perhaps most notoriously in the case of the EPA's report on dioxin. Indeed, *Science and Decisions* acknowledges that for some well-known problems, risk assessments should be bypassed altogether to focus on risk management. When the problem is obvious, Brewer notes, such as coal power plant emissions' negative impacts on health, the solution will not be found through further analytical assessment because the problem of what to do is "emotional, intentional, and political."[4]

Recommended improvements of the technical aspects of risk assessment are directed at how a problem is defined and assessed—is the right question being asked?—and making visible the process's inevitable uncertainties, variabilities, and assumptions. Perhaps most controversially, the report recommends significant changes to dose-response assumptions and exposure-assessment practices. The Silver or Platinum Book, as *Science and Decisions* is commonly called, recommends the end of the a priori assumption about a threshold dose-response relationship for noncarcinogenic substances. For decades, the default assumption in risk assessment has been that carcinogens follow a linear dose-response and noncarcinogens a threshold dose-response, meaning that toxic effects diminish at some low dose. The report recognizes that for some substances, such as heavy metals, there is no threshold of effect for neurological endpoints. If no threshold exists, an existing safety standard (based on a threshold dose-response) may need to be lowered. Further, the report recommends a shift away from a single reference dose or safety level to a "risk-specific dose." This risk-specific dose would be accompanied by an assessment of the percentage of the population "expected to be above or below a defined acceptable risk with a specific degree of confidence," or the probability of harm.[5] The result would be to make more visible who is at risk and how certain that information is.

The second important recommendation in *Science and Decisions* calls for the use of cumulative exposure assessment in risk-assessment protocols. The committee broadly defines cumulative assessments as those that "include combined risks posed by aggregate exposure to multiple agents or stressors; aggregate exposures includes all routes, pathways, and sources of exposure to a given agent or stressor."[6] This means that when the FDA conducts a risk assessment for BPA, for example, it needs to consider all routes of exposures, not only the one under its regulatory oversight—

food. This might include dermal exposure to BPA from thermal receipts, for example. Not surprisingly, industry has met such bold recommendations with concern. Particularly disquieting are the proposals to end threshold dose-response assumptions and to integrate cumulative exposure assessment. In response to *Science and Decisions*, the chemical industry's nonprofit organization, the International Life Sciences Institute, coordinated a series of meetings aimed at interpreting or reinterpreting its findings.[7]

On the more technical side—advancing the evaluation of chemical risks—considerable time and money are being spent to develop computational techniques to predict toxicity and human exposure. A vision for this future in testing is outlined in a 2007 National Research Council report entitled *Toxicity Testing in the 21st Century*. Greater use of in vitro (cell-based) screening based on human cell lines is proposed, along with computational models of biological systems and pharmacokinetic (absorption and distribution of a compound) modeling of human exposure. The objective is to move further away from whole-animal (in vivo) model testing and extrapolation from animals to humans.[8] At the moment, animal testing, which examines effects within the complex biological processes of development, growth, and function, is the only means to assess the accuracy and predictability of cell-based screens. One significant challenge to making this vision a reality is the central predicament in the endocrine-disruptor debate: What changes or perturbations to biological systems will be deemed adverse and predictive of disease development? In the process of reaching this objective, there will be a political and economic struggle to set the boundaries of what is defined as an adverse effect and what is not, which tests are appropriate and which are not.[9] For this reason, strong and specific conflict-of-interest policies are needed in this process.

As part of this vision, the EPA, along with other government institutions, is developing a number of methods and tools that will assess the predictive capabilities of cell-based screens. This involves aggregating existing historical toxicity data and comparing these outcomes with emerging data from automated, cell-based tests or screens. The Aggregated Computational Toxicology Resource and the Toxicity Reference Database accumulate vast bodies of data from research conducted over many decades. The Toxicity Forecaster, or ToxCast, collects data generated from 650 automated chemical screening tests. The screening data are integrated and used to generate a pictorial profile of a compound as part of the Toxicological Priority Index, or ToxPi.[10] It may take many years before this information will be used by regulators or, for that matter, by judges.

Outside of the regulatory and testing environment, nongovernmental organizations and private companies have initiated assessments tools of their own to improve decision making about chemicals in production and in products. For example, the small nonprofit Clean Production Action developed a tool called the Green Screen to assist companies such as Apple and Hewlett-Packard in evaluating chemicals used in their products. The screening tool allows companies to shift toward the use of safer compounds and anticipate phaseouts or bans. Lauren Heine, who designed the Green Screen, created a similar tool, called "cleangredients," for the cleaning products industry, managed by the nonprofit Green Blue Institute. The intention behind these tools is to allow manufacturers to make comparative assessments about possible safer substitutions based on existing available information. For example, on the basis of an assessment by a toxicity profiler, someone certified to gather the available data, Green Screen puts chemicals into four categories: red ("avoid"), orange ("use but search for safer substitutes"), yellow ("use but continue to look to improve"), and green ("prefer—safer chemical"). If a chemical were listed by the International Agency for Cancer Research as a probable human carcinogen, it would be considered a "very high hazard" and placed in the red category, chemicals to avoid. The Green Screen looks at the hazard qualities of chemicals independent of their use or potential exposure. The goal is to shift the market, step by step, toward preferred chemicals and, in so doing, to increase demand for safer chemicals and concomitantly investment in green chemistry.[11]

While these tools provide stepwise guidance for moving beyond well-defined, highly hazardous chemicals, they do not claim to set the boundaries for or definition of green chemistry. As with green or sustainable architecture, green chemistry aims to transform society through innovation in design—with design occurring at the molecular level. When many of the plastics on the market today were first developed in the 1930s and 1940s, health effects and environmental impacts were not considered in the performance and attributes of a chemical. Indeed, PhD programs for chemists to this day do not include training in toxicology, physiology, ecology, or environmental fate and transport.[12] Green chemistry integrates principles of renewable feedstocks (sources of chemicals), renewable energy, and reduction in hazardous properties. But for the chemists at the cutting edge of this initiative within the discipline, the biggest obstacle is in determining how to use high-throughput screens and in vivo assays and integrate them into the process of making "greener" molecules. As one green chemist pointedly asked at a meeting in the field in 2010, "What do we mean by *safe?*"[13]

Today, new computational tools, information databases, predictive models, and principles of green chemistry all seek to promote economic production while protecting public health—both are aspects of clean, sustainable production. In the case of BPA, determining safer substitutes is critical for reducing the potential risks this compound presents not just to a given individual but to the overall population, which is composed of many different vulnerable subpopulations. This cannot be simply achieved by a ban in the absence of larger science policy changes. There must be regulatory incentives and motivation that spur market competition for safer chemicals. Evaluating what is a safer substance will need to be considered in the context of what is known about the risks of BPA and will need to include independent evaluations that take into account low-level exposures, critical windows of exposure, and complex ways in which exposure might increase the risk of disease development at the level of the population. In the end, all of the research and the time and money spent understanding this single chemical may provide a pathway for navigating this complex new world of risk.

But the question "What do we mean by *safe?*," whether posed by a green chemist or the lay consumer, cannot be answered with new tests or the best interactive Web-based tool alone. Recognizing that safety is a highly negotiated concept shaped by what we know, how we know it, and who participates in the interpretation of that knowledge demands careful attention to implications of power, ideology, and values. As such, efforts to better manage financial conflicts of interest and bias, to make decision-making processes increasingly transparent, and to use the best available research hold the promise of saf*er* products and chemicals and healthier populations within a democratic society.

Notes

ACKNOWLEDGMENTS

1. Daniel Carpenter, *Reputation and Power: Organizational Image and Pharmaceutical Regulation at the FDA* (Princeton: Princeton University Press, 2010), 11.

INTRODUCTION

1. Heckman was once referred to as "Your Worship" by a member of the Society of the Plastics Industry (SPI). Thomas J. McGraw to members of the Plastics Bottle Division, SPI, confidential memorandum, February 1973, Hagley Museum and Archives, Soda House, Society of the Plastics Industry, Inc., Records 1936–87, corporate records (hereafter referred to as SPI papers).

2. Heckman recounts this story of the polystyrene in the chocolate mints in "A Plastics Industry Life," an unpublished white paper that he wrote in 2001 on the history of his career and provided to the author.

3. Styrene is listed as "possibly carcinogenic to humans" by the International Agency for Cancer Research. Most recently, the U.S. National Toxicology Program listed styrene as "reasonably anticipated to be a human carcinogen" in the twelfth edition of its *Report on Carcinogens*. See International Agency for Cancer Research, *Some Traditional Herbal Medicines, Some Mycotoxins, Naphthalane and Styrene*, IARC Monograph on the Evaluation of Carcinogenic Risks to Humans 82 (Lyon: IARC Press/Geneva: World Health Organization, 2002); National Toxicology Program, "Styrene," in *Report on Carcinogens*, 12th ed. (Research Triangle Park, NC: U.S. Department of Health and Human Services, 2011).

4. Heckman, "Plastics Industry Life."

5. For general history on the plastics industry, see Susan Freinkel, *Plastics: A Toxic Love Story* (New York: Houghton Mifflin Harcourt, 2011); and Jeffrey Meikle, *American Plastics: A Cultural History* (New Brunswick: Rutgers University Press, 1997). See also John R. Lawrence, director and technical liaison,

Society of the Plastics Industry, to Kenneth Burgess, Dow Chemical Co., June 21, 1973, executive officers' correspondences, SPI papers; Rudolph Unger, "Chemists Challenged to Make Plastics Safe," *Chicago Tribune*, November 17, 1977, N8.

6. Heckman, "Plastics Industry Life," 17.

7. Alfred Chandler Jr., *Shaping the Industrial Century: The Remarkable Story of the Evolution of the Modern Chemical and Pharmaceutical Industries* (Cambridge, MA: Harvard University Press, 2005); Ulrich Beck, *Risk Society: Towards a New Modernity* (Newbury Park, CA: Sage Publications, 1992).

8. Ulrich Beck, *World at Risk* (Cambridge: Polity Press, 2007), 25.

9. A.M. Calafat et al., "Urinary Concentrations of Bisphenol A and 4-Nonylphenol in a Human Reference Population," *Environmental Health Perspectives* 113 (2005): 391–95; L.N. Vandenberg et al., "Human Exposure to Bisphenol A (BPA)," *Reproductive Toxicology* 24 (2007): 139–77.

10. Daniel Carpenter, *Reputation and Power: Organizational Image and Pharmaceutical Regulation at the FDA* (Princeton: Princeton University Press, 2010), 11.

11. Ibid., ch. 1, "Reputation and Regulatory Power," 33–70, provides a detailed discussion of the relationship between reputation and power and the role of various audiences, such as industry, in establishing or eroding authority.

12. Jerome H. Heckman, "The New Pressure for Associations to Act in the Public Interest," *Association Management*, February 1975, 28–31.

13. Jerome Heckman, Heckman and Keller, LLC, interview by author, July 15, 2010.

14. Christopher C. Sellers, *Hazards of the Job: From Industrial Disease to Environmental Health Science* (Chapel Hill: University of North Carolina Press, 1997).

15. E. Stokstad, "Biomonitoring: A Snapshot of the U.S. Chemical Burden," *Science* 304 (2004): 1893; E. Stokstad, "Biomonitoring: Pollution Gets Personal," *Science* 304 (2004): 1892–94.

16. Mark Kirschrer, "Chemical Profile: Bisphenol-A," *Chemical Market Reporter* 266 (2004): 27.

17. E.C. Dodds and W. Lawson, "Synthetic Oestrogenic Agents without the Phenanthrene Nucleus," *Nature* 137 (June 13, 1936): 996; E.C. Dodds et al., "Oestrogenic Activity of Certain Synthetic Compounds," *Nature* 141 (1938): 247–48; F. Dickens, "Edward Charles Dodds. 13 October 1899–16 December 1973," *Biographical Memoirs of Fellows of the Royal Academy* 21 (1975): 241.

18. Elizabeth Siegel Watkins, *The Estrogen Elixir: A History of Hormone Replacement Therapy in America* (Baltimore: Johns Hopkins University Press, 2007).

19. Ibid., 1.

20. Dolores Iberreta and Shanna H. Swan, "The DES Story: Long-Term Consequences of Prenatal Exposure," in *Late Lessons from Early Warnings:*

The Precautionary Principle, 1896–2000, ed. Poul Harremoës et al., European Environment Agency Environmental Issue Report 2 (Luxembourg: Office for Official Publications of the European Communities, 2001), www.eea.europa .eu/publications/environmental_issue_report_2001_22.

21. Theo Colborn, Dianne Dumanoski, and John Peterson Myers, *Our Stolen Future: Are We Threatening Our Fertility, Intelligence, and Survival? A Scientific Detective Story* (New York: Penguin Books, 1996).

22. Sheldon Krimsky, *Hormonal Chaos: The Scientific and Social Origins of the Environmental Endocrine Hypothesis* (Baltimore: Johns Hopkins University Press, 2000).

23. Nancy Langston, *Toxic Bodies: Hormone Disruptors and the Legacy of DES* (New Haven: Yale University Press, 2010).

24. Iberreta and Swan, "DES Story."

25. Langston, *Toxic Bodies*, xiii.

1. PLASTIC FOOD

1. Testimony of L. G. Cox, Beech-Nut Packing Co., in Hearings before the House Select Comm. to Investigate the Use of Chemicals in Food Products, 82nd Cong., 2nd sess., 1952 (hereafter referred to as the Chemicals in Food Hearings 1952), pt. 2, January 31, 1952, 1388.

2. Ibid.

3. Ibid.

4. Daniel Yergin, *The Prize: The Epic Quest for Oil, Money, and Power* (New York: Simon and Schuster, 1991). At the foundation of this new order was the readjustment of oil "rents," the distribution of profits between oil corporations and the major oil-producing nations. Oil-producing countries began to demand a greater share of profits from oil corporations. In the 1940s, Venezuela, followed by Saudi Arabia and Kuwait in the 1950s, adopted the fifty-fifty principle, which held that the profits of an oil company should be split with the oil-producing country through the collection of taxes and royalty fees. As part of these agreements, U.S. oil companies were exempted from paying U.S. income tax, thereby diverting this money to the treasuries of Middle Eastern countries and Venezuela. The fifty-fifty principle was a precarious balancing act among global power, oil, and wealth. To preserve the balance and the concept that the agreement was fair, U.S. foreign policy in the Middle East and Latin America centered on maintaining foreign governments that were "friendly" to the principle. For the next two decades, this balance held steady, and oil remained cheap and abundant, providing the necessary inputs for the petrochemical revolution.

5. Ibid., 541.

6. Ibid., 409.

7. Louis Galambos and Joseph A. Pratt, *The Rise of the Corporate Commonwealth: U.S. Business and Public Policy in the Twentieth Century* (New York: Basic Books, 1988).

8. Alfred Chandler Jr., *Shaping the Industrial Century: The Remarkable Story of the Evolution of the Modern Chemical and Pharmaceutical Industries* (Cambridge, MA: Harvard University Press, 2005), 27.

9. Peter Spitz, *Petrochemicals: The Rise of an Industry* (New York: John Wiley and Sons, 1988), xiii.

10. Chandler, *Shaping the Industrial Century*, 148–50; Industrial Development Organization of the United Nations, "The Petrochemical Industry," in *Perspectives for Industrial Development in the Second United Nations Development Decade* (New York: United Nations, 1973), 1.

11. Thomas R. Dunlap, *DDT: Scientists, Citizens, and Public Policy* (Princeton: Princeton University Press, 1981).

12. "Reynolds Cuts Plastic Film," *New York Times*, May 7, 1949, 20; "News of Food," *New York Times*, May 18, 1948, 20; "Outlook 'Normal' in Packing Trade," *New York Times*, June 26, 1949, 6; William Enright, "Tinless Cans Use Plastic Coatings," *New York Times*, August 11, 1940, 49.

13. Winston Williams, "James J. Delaney, 86, a Democrat and Former Queens Congressman," *New York Times*, May 25, 1987.

14. Judy Gardner, "Consumer Report/Debate Intensifies over Law Banning Cancer-Causing Food Additives," *National Journal*, September 30, 1972, 1534–43.

15. "Food Chemicals Seen Raising Big Problem," *New York Times*, January 4, 1951, 47; "Food ManDepicts Fight on Pesticides," *New York Times*, February 1, 1952, 23.

16. Maurice H. Seevers, "Perspective versus Caprice in Evaluating Toxicity of Chemicals in Man," *Journal of the American Medical Association* 153 (1953): 1329–33.

17. James Delaney, "Peril on Your Food Shelf," *American Magazine*, July 1951, 112.

18. "Terror Campaign Laid to Politicians," *New York Times*, February 20, 1952, 45.

19. Lars Noah and Richard A. Merrill, "Starting from Scratch? Reinventing the Food Additive Approval Process," *Boston University Law Review* 78 (1998): 333–36.

20. Marc Law, "History of Food and Drug Regulation in the United States," EH.Net Encyclopedia, ed. Robert Whaples, October 11, 2004, http://eh.net/encyclopedia/article/Law.Food.and.Drug.Regulation.

21. Daniel Carpenter, *Reputation and Power: Organizational Image and Pharmaceutical Regulation at the FDA* (Princeton: Princeton University Press, 2010), 74–75.

22. Noah and Merrill, "Starting from Scratch?," n. 16. Noah and Merrill note that the agency "frequently could not satisfy this burden even if it had some legitimate basis for concern about safety."

23. Chemicals in Food Hearings 1950, 2.

24. The congressional members on the committee included Delaney, who served as chairman, Thomas Abernethy (D-MS), Erland H. Hedrick (D-WV),

Paul Jones (D-MO), Frank Keefe (R-WI), Arthur L. Miller (R-NE), and Gordon McDonough (R-CA). The pure food activists included several of the original promoters of the Pure Food and Drug Act of 1906, one of whom was Leslie B. Wright, chairman, Department of Legislation, General Federation of Women's Clubs. This organization, which represented five million women in 1950, had urged the passage of food safety laws since 1904. In the early 1950s, the organization supported a new law that would restrict any "nonnutritive ingredients" unless an additive was useful and shown to be safe. Statement of Mrs. Leslie B. Wright, chairman, Dept. of Legislation, General Federation of Women's Clubs, in Chemicals in Food Hearings 1950, pt. 1, December 15, 1950, 775–76.

25. Statement of John Foulgor on behalf of the Manufacturing Chemists' Association [MCA], in Chemicals in Food Hearings 1950, pt. 1, December 1, 1950, 501–19.

26. Kehoe's connection with the lead industry is detailed in Gerald E. Markowitz and David Rosner, *Deceit and Denial: The Deadly Politics of Industrial Pollution* (Berkeley: University of California Press, 2002), 47.

27. Testimony of Dr. Robert A. Kehoe, director of the Kettering Laboratory, College of Medicine, University of Cincinnati, in Chemicals in Food Hearings 1950, pt. 1, December 13, 1950, 745–62.

28. Testimony of C. W. Crawford, commissioner of the FDA, in Chemicals in Food Hearings 1950, pt. 1, November 28, 1950, 337–50.

29. Blanche Wiesen Cook, *Eleanor Roosevelt*, vol. 2 (New York: Viking, 1992), 84.

30. Testimony of C. W. Crawford, commissioner of the FDA, in Chemicals in Food Hearings 1950, pt. 1, November 28, 1950, 337–50.

31. Testimony of Wilhelm C. Hueper, in *Hearings before the House Select Comm. to Investigate the Use of Chemicals in Food Products*, 82nd Cong., 2nd sess., 1952 (hereafter referred to as the Chemicals in Food Hearings 1952), pt. 2, January 29, 1952, 1353–82.

32. Robert Proctor, *Cancer Wars: How Politics Shapes What We Know and Don't Know about Cancer* (New York: Basic Books, 1995), 45.

33. Wilhelm Hueper, *Occupational Tumors and Allied Diseases* (Springfield, IL: Charles C. Thomas, 1942); Proctor, *Cancer Wars*, 45.

34. Rachel Carson, *Silent Spring* (1962; repr., Boston: Houghton Mifflin, 1994), 222; Linda Lear, *Rachel Carson: Witness for Nature* (New York: Henry Holt, 1997).

35. Wilhelm Hueper, chief, Environmental Cancer Section, to Dr. John R. Heller, director, National Cancer Institute [NCI], memorandum, June 8, 1959, National Library of Medicine Archives, History of Medicine, Wilhelm Hueper Papers 1920–1981, Series 1: Autobiographical Information, MS C 341, Box 1, Folder: Autobiography—Draft, 101–50.

36. Proctor, *Cancer Wars*, 45.

37. Associated Press, "U.S. Paper on Cancer Spurs Secrecy Probe," *Evening Star*, April 10, 1958, B8.

38. Statement of Dr. William E. Smith, Englewood, NJ, July 18 and July 19 (resumed), 1957, in *Hearings before a Subcomm. of the Comm. on Interstate and Foreign Commerce, on Bills to Amend the Federal Food, Drug, and Cosmetic Act with Respect to Chemical Additives in Food, House of Reps.*, 85th Cong., 1st and 2nd sess., 1957–58 (hereafter referred to as FFDCA Hearings 1957–58).

39. Dr. William Smith's testimony in Charles Wesley Dunn, ed., *Legislative Record of 1958 Food Additives Amendment to Federal Food, Drug, and Cosmetic Act*, Food Law Institute Series (New York: Commerce Clearing House, 1958), 83–84.

40. E. C. Dodds et al., "Oestrogenic Activity of Certain Synthetic Compounds," *Nature* 141 (1938): 247–48; F. Dickens, "Edward Charles Dodds: 13 October 1899–16 December 1973," *Biographical Memoirs of Fellows of the Royal Academy* 21 (1975): 227–67. For histories of DES, see Nancy Langston, *Toxic Bodies: Hormone Disruptors and the Legacy of DES* (New Haven: Yale University Press, 2010), xiii; Alan I. Marcus, *Cancer from Beef: DES, Federal Food Regulation, and Consumer Confidence* (Baltimore: Johns Hopkins University Press, 1994); Elizabeth Siegel Watkins, *The Estrogen Elixir: A History of Hormone Replacement Therapy in America* (Baltimore: Johns Hopkins University Press, 2007); Susan E. Bell, "Gendered Medical Science: Producing a Drug for Women," *Feminist Studies* 21, no. 3 (1995): 469–500; and Nelly Oudshoorn, *Beyond the Natural Body* (New York: Routledge, 1994).

41. Watkins, *Estrogen Elixir*, 13–14; Oudshoorn, *Beyond the Natural Body*, 28.

42. Dickens, "Edward Charles Dodds," 241.

43. Ibid.; E. C. Dodds and W. Lawson, "Synthetic Oestrogenic Agents without the Phenanthrene Nucleus," *Nature* (June 13, 1936): 996.

44. Dickens, "Edward Charles Dodds." Dickens cites a number of collaborative papers by Cook and Dodds, including J. W. Cook and E. C. Dodds, "Sex Hormones and Cancer-Producing Compounds," *Nature* (London) 131 (1933): 205; J. W. Cook, E. C. Dodds, and C. L. Hewett, "A Synthetic Oestrus-Exciting Compound," *Nature* (London) 131 (1933): 56–57; and J. W. Cook et al., "The Oestrogenic Activity of Some Condensed Carbon Ring Compounds in Relation to Their Other Biological Activity," *Proceedings of the Royal Society of London B* 114 (1934): 272–86.

45. Dodds and Lawson, "Synthetic Oestrogenic Agents."

46. Dickens, "Edward Charles Dodds."

47. Ibid.

48. Langston, *Toxic Bodies*, 32–44; Watkins, *Estrogen Elixir*, 28–29.

49. Langston, *Toxic Bodies*, 59. Langston and others have explored how gendered assumptions about women's health, the medicalization of the female body, and pathologizing of female reproductive events (e.g., hot flashes, miscarriage, menstruation) informed the development of estrogen therapies. Bell, "Gendered Medical Science"; Oudshoorn, *Beyond the Natural Body*.

50. Langston, *Toxic Bodies*, 69–71; Marcus, *Cancer from Beef*.

51. Granville F. Knight et al., "Possible Cancer Hazard Present by Feeding Diethylstilbestrol to Cattle," statement presented at the Symposium on Medicated Feeds, conducted by the U.S. Dept. of Health, Education and Welfare, FDA, Veterinary Branch, Washington, DC, January 24, 1956, in uncatalogued collection of M. E. Grenander Department of Special Collections and Archives, State University of New York at Albany (hereafter referred to as Delaney papers).

52. Ibid.

53. President Dwight D. Eisenhower quoted in Marcus, *Cancer from Beef*, 35–36.

54. Statement of Dr. W. C. Hueper, NCI, NIH, Bethesda, MD, in FFDCA Hearings 1957–58, August 7, 1957, 370.

55. Testimony of Dr. Anton Julius Carson, professor emeritus, University of Chicago Medical School, in Chemicals in Food Hearings 1950, pt. 1, September 14, 1950, 9.

56. Statement of Don D. Irish, director of biochemical research, Dow Chemical Company, in Chemicals in Food Hearings 1950, pt. 1, December 1, 1950, 468–73.

57. Testimony of Dr. Robert Harris, Nutritional Biochemistry Laboratories, Dept. of Food Technology, Massachusetts Institute of Technology, in Chemicals in Food Hearings 1950, pt. 1, December 1, 1950, 473–89. The duration of the tests referred to the amount of time a plastic wrap was in contact with meat during the testing of migrating levels.

58. Wilhelm Hueper, "Polyvinylpyrrolidone: Canceritenic Agents for Rats," *Proceedings of the American Journal of Cancer Research* 2 (1956): 120; L. M. Lusky and A. A. Nelson, "Fibrosarcomas Induced by Multiple Subcutaneous Injections of Carboxymethyl Cellulose (CMC), Polyvinylpyrrolidone (PVP) and Polyoxyethylene Sorbitan Monostearate (TWEEN 60)," *Federal Proceedings* 16 (1957): 318.

59. B. S. Oppenheimer, E. T. Oppenheimer, and A. P. Stout, "Sarcomas Induced in Rodents by Imbedding Various Plastic Films," *Proceedings of the Society for Experimental Biology and Medicine* 79, no. 3 (1952): 366–69.

60. B. S. Oppenheimer et al., "Further Studies of Polymers as Carcinogenic Agents in Animals," *Cancer Research* 15, no. 5 (1955): 333–40.

61. A. F. Fitzhugh et al., "Malignant Tumors and High Polymers," *Science* 118, no. 3078 (1953): 783–84; B. S. Oppenheimer, E. T. Oppenheimer, and A. P. Stout, "Carcinogenic Effect of Imbedding Various Plastic Films in Rats and Mice," *Surgical Forum* 4 (1953): 672–76; B. S. Oppenheimer et al., "Malignant Tumors Resulting from Embedding Plastics in Rodents," *Science* 118, no. 3063 (1953): 305–06; B. S. Oppenheimer et al., "The Latent Period in Carcinogenesis by Plastics in Rats and Its Relation to the Presarcomatous Stage," *Cancer* 11, no. 1 (1958): 204–13; B. S. Oppenheimer et al., "Studies of the Mechanism of Carcinogenesis by Plastic Films," *Acta Unio Internationalis Contra Cancrum* 15 (1959): 659–63.

62. W.K. *[sic]*, "Notes on Science: Plastic Film Linked to Cancer," *New York Times*, May 11, 1952, E11.

63. A list of known chemicals used in food production was included in Chemicals in Food Hearings 1950, pt. 1, 69–75.

64. "Food Chemicals," 47; Lucia Brown, "Postwar Use of Chemicals in Food Dubbed 'Alarming,'" *Washington Post*, January 15, 1951, B3.

65. Original copies of letters from all around the country can be found in the Delaney papers.

66. "Congressman Says Actress's Speech Helped Bar Cyclamates," *New York Times*, October 22, 1969, 26.

67. George Merck quoted in "MCA Raises $170,000 for Publicity, Set Up Mobilization Committee," *Chemical and Engineering News* 29 (1951): 5442.

68. Jerome H. Heckman, "The New Pressure for Associations to Act in the Public Interest," *Association Management*, February 1975, 31.

69. Statement of Dr. William E. Smith, in FFDCA Hearings 1957–58, July 18, 1957, 172.

70. Letter from William Smith to James Delaney, extension of remarks of Hon. James J. Delaney, "Chemical Additives in Our Food Supply Can Cause Cancer," 85th Cong., 1st sess., *Congressional Record* 103, pt. 2 (February 21, 1957): Appendix, in Delaney papers.

71. Food Packaging Materials Committee of the Society of the Plastics Industry, meeting minutes, June 5, 1957, Hagley Museum and Archives, Soda House, Society of the Plastics Industry [SPI], Inc., Records 1936–1987, Committee Records (hereafter referred to as SPI papers).

72. Statement of George Larrick, FDA commissioner, in *Federal Food, Drug, and Cosmetic Act (Chemical Additives in Food): Hearings before the House Subcommittee of the Committee on Interstate and Foreign Commerce, House of Representatives, on H.R. 4475, H.R. 7605, 7606, 8748, 7607, 7764, 8271, 8275*, 84th Cong., 2nd sess., 1956 (hereafter cited as FFDCA Hearings 1956), February 14, 1956, 193.

73. Ibid.

74. Carpenter, *Reputation and Power*, 119.

75. Statement of George Larrick, FDA commissioner, in FFDCA Hearings 1956, February 14, 1956, 193.

76. Statement of Lawrence Coleman on behalf of theMCA, in FFDCA Hearings 1957–58, July 17, 1957, 114.

77. "Food Additives Arouse Dispute," *New York Times*, June 30, 1957, 34.

78. Robert C. Barnard, "Some Regulatory Definitions of Risk: Interaction of Scientific and Legal Principles," *Regulatory Toxicology and Pharmacology* 11 (1990): 201–11.

79. Sec. 201 (s) [21 USC § 321], Federal Food, Drug and Cosmetic Act.

80. Sec. 201 (s) [21 USC § 321], Federal Food, Drug and Cosmetic Act. The phrase "generally recognized, among experts" derived from the 1938 law guiding drug regulation. In 1937, a drug company dissolved sulfanilamide in a new solvent, and the new combination resulted in the deaths of hundreds of

people, including children. The FDA responded to this tragedy by defining a new drug as one not yet recognized among experts as safe. Such a term placed the authority for determining safety in the hands of not only experts but more specifically regulatory agency experts.

81. Commissioner Larrick failed to win support for the agency's authority to determine the "functional value" of a chemical. This would have created regulatory oversight of the benefits of a chemical and its impact on nutrition. Statement of George Larrick, FDA commissioner, in FFDCA Hearings 1956, February 14, 1956, 205.

82. The Walter-Logan bill, introduced by the American Bar Association in 1940, became the Administrative Procedures Act (APA) when signed into law six years later. For a discussion of the APA and the debates leading to its passage, see George B. Shepherd, "Fierce Compromise: The Administrative Procedures Act Emerges from New Deal Politics," *Northwestern University Law Review* 90 (1996): 1557.

83. *Universal Camera Corporation v. National Labor Relations Board*, 340 U.S. 474 (1951); Court decision discussed in Cass Sunstein, "Constitutionalism after the New Deal," *Harvard Law Review* 101, no. 2 (1987): 421–510.

84. Shepherd, "Fierce Compromise."

85. Statement of William Goodrich, assistant general counsel, FDA, in FFDCA Hearings 1956, February 14, 1956, 221.

86. Roy Newton, vice president, Swift and Co., to Vincent Kleinfeld, chief counsel, Committee to Investigate the Use of Chemicals in Food Products, in Chemicals in Food Hearings 1952, pt. 2, January 22, 1952, 1407–10.

87. Dies's words can be found in the statement of George Larrick, FDA commissioner, in FFDCA Hearings 1956, February 14, 1956, 205.

88. Statement of Dr. William E. Smith, in FFDCA Hearings 1957–58, July 18, 1957, 172.

89. The Delaney clause appears in Sec. 409 [21 USC § 348] of the Federal Food, Drug and Cosmetic Act.

90. The committee discussed the following two studies: Edgar A. Bering Jr. et al., "The Production of Tumors in Rats by the Implantation of Pure Polyethylene," *Cancer Research* 15, no. 5 (1955): 300–301; and B. Oppenheimer, Oppenheimer, and Stout, "Sarcomas Induced in Rodents." MCA, Plastics Committee, meeting minutes, February 11, 1957, SPI papers.

91. MCA, Plastics Committee, meeting minutes, February 11, 1957, SPI papers.

92. Rex H. Wilson and William E. McCormick, "Plastics: The Toxicology of Synthetic Resins," *AMA Archives of Industrial Health* 21 (1960): 56–68.

93. Jerome Heckman, "A Plastics Industry Life," unpublished white paper, 2001, supplied to the author; Jerome Heckman, interview by author, July 14, 2010.

94. Heckman was added to the Plastics Hall of Fame in 2004; see "Jerome H. Heckman," March 29, 2004, www.plasticsacademy.org/swp/articles

.php?articleId=71. The Plastics Hall of Fame was created in 1972 by the industry journal *Modern Plastics* and the SPI.

95. Report from Jerome Heckman, Food Packaging Materials Committee of the SPI, meeting minutes, December 16, 1958, committee records, SPI papers.

96. Jerome Heckman, interview by author, July 15, 2010.

97. Technical Information Subcommittee of the Food Packaging Materials Committee of the SPI, meeting minutes, March 9, 1960, committee records, SPI papers.

98. The statistic on polyethylene was in a letter from J. K. Kirk of the FDA, discussed by Heckman at a meeting of the Food Packaging Materials Committee of the Society of the Plastics Industry, meeting minutes, June 23, 1960, committee records, SPI papers.

99. Heckman's remarks are in the meeting minutes of the Food Packaging Materials Committee of the SPI, December 1, 1960, 2, committee records, SPI papers.

100. Ibid., 6.

2. THE "TOXICITY CRISIS" OF THE 1960S AND 1970S

1. Victor Cohn, "Plastics Residues Found in Bloodstreams," *Washington Post,* January 18, 1972, A3.

2. Daniel Carpenter, *Reputation and Power: Organizational Image and Pharmaceutical Regulation at the FDA* (Princeton: Princeton University Press, 2010), 119.

3. Linda Lear, *Rachel Carson: Witness for Nature* (New York: Henry Holt, 1997), 425.

4. Ibid., 428.

5. Dwight D. Eisenhower, "Farewell Radio and Television Address to the American People," January 17, 1961, Public Papers of the Presidents, www.presidency.ucsb.edu/ws/index.php?pid=12086&st=&st1=#axzz1uUMOzUSE.

6. Lear, *Rachel Carson,* 440.

7. Barbara Resnick Troetel, *Three-Part Disharmony: The Transformation of the Food and Drug Administration in the 1970s,* pub. no. 9630515 (Ann Arbor: Proquest/UMI, 1996), 83–90.

8. Robert Proctor, *Cancer Wars: How Politics Shapes What We Know and Don't Know about Cancer* (New York: Basic Books, 1995).

9. Jerome Heckman, report to the Food Packaging Materials Committee of the Society of the Plastics Industry [SPI], meeting minutes, December 16, 1958, Hagley Museum and Archives, Soda House, SPI, Inc., Records 1936–1987, Committee Records (hereafter referred to as SPI papers).

10. Nancy Langston, *Toxic Bodies: Hormone Disruptors and the Legacy of DES* (New Haven: Yale University Press, 2010), 59.

11. Mel DeMunn, Farm Report, U.S.A., to Rep. John D. Dingell, June 23, 1960, in uncatalogued collection of M. E. Grenander Department of Special

Collections and Archives, State University of New York at Albany (hereafter referred to as Delaney papers).

12. Allan J. Greene, administrative vice president, Chas. Pfizer and Co, Inc. Legal Division, to Hon. Oren Harris, Chairman of Interstate and Foreign Commerce Committee, March 16, 1960, Delaney papers.

13. Edward Harris to Rep. John D. Dingell, February 10, 1960, Delaney papers. Delaney's archive included correspondence between Wilhelm Hueper and Edward Harris and between Delaney and Hueper. Abbott Laboratories was a producer of DES.

14. Joseph V. Rodricks, "FDA's Ban of the Use of DES in Meat Production: A Case Study," *Agriculture and Human Values* 3 (Winter–Spring 1986): 10–25.

15. Rachel Carson, *Silent Spring* (1962; repr., Boston: Houghton Mifflin, 1994), 222.

16. Ibid., 127.

17. Harold Stewart, consultant to the National Cancer Institute [NCI], to Wilhelm Hueper, June 13, 1973, National Library of Medicine Archives, History of Medicine, Wilhelm Hueper Papers 1920–1981 (hereafter referred to as Hueper papers), Box 3, Folder: Correspondence 1966–1974.

18. G. C. Thelen, "FDA Shakeup Demotes Critical Pathologist," *Washington Star*, May 26, 1970, Delaney papers.

19. James Delaney to Wilhelm Hueper, August 13, 1970, Delaney papers; Thelen, "FDA Shakeup."

20. An undated, unauthored, and untitled white paper in the Delaney papers details the contents of the memo. No original copy of the memo was found in Delaney's papers.

21. National Academy of Sciences [NAS], National Academy of Engineering, National Research Council [NRC], *Annual Report, Fiscal Year 1968–69* (Washington, DC, 1972), 130; Troetel, *Three-Part Disharmony*, 31.

22. "Finch Raps FDA on Sweeteners," *Washington Post*, October 8, 1969, A6; Jacqueline Verrett, interview, *NBC Nightly News*, October 1, 1969, quoted in Troetel, *Three-Part Disharmony*, 28.

23. Louise Hutchinson, "Artificial Sweetener Ban to Be Ordered," *Chicago Tribune*, October 18, 1969, W1.

24. Richard A. Merrill and Michael R. Taylor, "Saccharin: A Case Study of Government Regulation of Environmental Carcinogens," *Agriculture and Human Values* 3 (Winter–Spring 1986): 33–72.

25. "New Broom at FDA," *Science News*, January 31, 1970, 120–21.

26. James Delaney to Wilhelm Hueper, August 13, 1970, Delaney papers; Thelen, "FDA Shakeup"; Nicholas Wade, "Delaney Anti-Cancer Clause: Scientists Debate on Article of Faith," *Science* 177 (1972): 588–91.

27. "Delaney Demands Probe of the FDA," *Long Island Press*, June 5, 1970, 3; James Delaney to FDA commissioner Charles Edwards, June 24, 1971, Delaney papers.

28. Food and Drug Administration, "Charles Edwards, M.D.," June 4, 2009, www.fda.gov/AboutFDA/CommissionersPage/PastCommissioners/ucm113436.htm.

29. Gerald E. Markowitz and David Rosner, *Deceit and Denial: The Deadly Politics of Industrial Pollution* (Berkeley: University of California Press, 2002), 151–53.

30. Charles C. Edwards quoted in Troetel, *Three-Part Disharmony*, 74.

31. Wade, "Delaney Anti-Cancer Clause," 589.

32. James Delaney to FDA commissioner Charles Edwards, June 24, 1971, Delaney papers.

33. Charles C. Edwards, MD, FDA commissioner, remarks presented at Temple University, Philadelphia, May 5, 1970, Delaney papers.

34. Merrill and Taylor, "Saccharin," 47.

35. Jean Carper, "Danger of Cancer in Food," *Saturday Review*, September 5, 1970, 47–57, Delaney papers.

36. Wade, "Delaney Anti-Cancer Clause," 590. According to Wade's article, the report (see next note) was printed but never published because of the objections, based on "literary style, not principle," of an NAS committee member.

37. Umberto Saffiotti, *Ad Hoc Committee on the Evaluation of Low Levels of Environmental Chemical Carcinogens* (Bethesda, MD: NCI, 1970); Wade, "Delaney Anti-Cancer Clause," 590.

38. William J. Darby, chairman, Food Protection Committee, NRC, Food and Nutrition Board, *Guidelines for Estimating Toxicologically Insignificant Levels of Chemicals in Food,* pamphlet (Washington, DC: NAS, NRC, 1969); Wilhelm Hueper to James Delaney, August 29, 1970; Samuel Epstein, "The Delaney Clause," *Preventative Medicine* 2, no. 1 (1973): 140–49; and Health Research Group, *Cancer Prevention and the Delaney Clause*, pamphlet (Washington, DC: Health Research Group, 1977), all in Delaney papers.

39. Wade, "Delaney Anti-Cancer Clause."

40. Elise Jerald to James Delaney, July 10, 1972, Delaney papers.

41. A. L. Herbst, H. Ulfelder, and D. C. Poskanzer, "Adenocarcinoma of the Vagina: Association of Maternal Stilbestrol Therapy with Tumor Appearance in Young Women," *New England Journal of Medicine* 284, no. 15 (1971): 878–81. For an overview of the Massachusetts General study, see Theo Colborn, Dianne Dumanoski, and John Peterson Myers, *Our Stolen Future: Are We Threatening Our Fertility, Intelligence, and Survival?: A Scientific Detective Story* (New York: Penguin Books, 1996), 54–55; and Sheldon Krimsky, *Hormonal Chaos: The Scientific and Social Origins of the Environmental Endocrine Hypothesis* (Baltimore: Johns Hopkins University Press, 2000), 10.

42. Elizabeth Siegel Watkins, *The Estrogen Elixir: A History of Hormone Replacement Therapy in America* (Baltimore: Johns Hopkins University Press, 2007), 26.

43. Alan I. Marcus, *Cancer from Beef: DES, Federal Food Regulation, and Consumer Confidence* (Baltimore: Johns Hopkins University Press, 1994), 117.

44. Charles C. Edwards quoted in Wade, "Delaney Anti-Cancer Clause," 588.

45. Rodricks, "FDA's Ban," 18; Peter Hutt, "Regulatory History of DES," *American Statistician* 36 (August 1982): 267; Merle Ellis, "Additive in Trouble Again," *Los Angeles Times*, February 10, 1977, K24; Don M. Larrimore, "U.N. Body Seeks Ban on Synthetic Hormone Used in Cattle Feed," *Washington Post*, November 11, 1971, F6; Richard D. Lyons, "In the Cause of Safety," *New York Times*, April 29, 1973, 274.

46. Hutt, "Regulatory History of DES."

47. Donald Kennedy, phone interview by author, October 22, 2010.

48. Clinton R. Miller, "Bill to Weaken Delaney Amendment Must Be Defeated," *National Health Federation Bulletin*, September 1972, Delaney papers. The bulletin provides background information on the National Health Federation, founded in 1955: "Its members believe that health freedoms are inherently guaranteed to us as human beings, and outright to them as Americans as implied in the words, 'life, liberty and the pursuit of happiness.' Yet, frequently, these freedoms and rights have been and continue to be violated. Too often, as a result of the unopposed pressures from organized medicine, the chemical industries, pharmaceutical manufacturers, and others, laws and regulations have been imposed which better serve these special-interests than the public at large." In the 1972 bulletin, the National Health Federation strongly supported Nelson's bills and encouraged its members to inform their congressmen that no amount of a carcinogen was safe.

49. Merrill and Taylor, "Saccharin," 50.

50. Rodricks, "FDA's Ban," 16.

51. FDA, "FDA Revises Definition of the Term 'No Residue,'" press release, December 24, 2002, www.fda.gov/AnimalVeterinary/NewsEvents/CVMUpdates/ucm125995.htm.

52. Chapter 1, Food and Drug Administration, Department of Health and Human Services, Code of Federal Regulations, Title 21, Part 500.84–88, www.access.gpo.gov/nara/cfr/waisidx_02/21cfr500_02.html.

53. Roland R. MacBride, "Toxicity: A Gnawing Concern That Could Get Tougher to Live With," *Modern Plastics* 57 (October 1973): 73.

54. Mike Causey, "Nader Study Group Runs into Delay," *Washington Post*, November 2, 1972, E11.

55. Barry Commoner, *The Closing Circle: Nature, Man, and Technology* (New York: Knopf, 1971), 174. Commoner's thesis was also in part an argument against the "limits to growth" thesis, which emphasized overpopulation as the source of environmental degradation. He considered calls for population control a form of political repression.

56. Ibid., 228–29; R. J. Jaeger and R. J. Rubin, "Plasticizers from Plastic Devices Extraction, Metabolism, and Accumulation by Biological Systems," *Science* 170 (October 23, 1970): 461.

57. Arthur J. Warner, Charles H. Parker, Bernard Baum, and DeBell & Richardson, Inc., *Plastics Solid Waste Disposal by Incineration or*

Landfill (Washington, DC: Manufacturing Chemists' Association [MCA], 1971), 18.

58. "Final Data from MRI Research Confirm Plastics' Environmental Superiority," *Modern Plastics* 52 (February 1975): 12.

59. Wilhelm Hueper, "Autobiography," Hueper papers, Box 1, Folder: Autobiography—Draft, 192.

60. Markowitz and Rosner, *Deceit and Denial*, 168–94.

61. Ibid.

62. David Burnham, "Ban Is Asked on Vinyl Chloride in Food Packages," *New York Times*, July 2, 1975, 18.

63. Herbert Stockinger quoted in David Burnham, "Official Assails Cancer Institute," *New York Times*, April 7, 1975, 5. On the establishment of the American Conference of Governmental Industrial Hygienists, see Proctor, *Cancer Wars*, 157.

64. Jerome Heckman quoted in Burnham, "Ban Is Asked."

65. "Energy and Environment: Plastics Curbs May Cost 1.6 Million Jobs," *Los Angeles Times*, June 21, 1974, OC2.

66. Nicholas Ashford, *Crisis in the Workplace: Occupational Disease and Injury: A Report to the Ford Foundation* (Cambridge, MA: MIT Press, 1976), 177.

67. David Vogel, "The Political Resurgence of Business," in *Fluctuating Fortunes: The Political Power of Business in America* (New York: Basic Books, 1989). Vogel argues that in the mid- to late 1970s business increased "its political effectiveness and reshape[d] the prevailing political and intellectual climate of opinion" (193). Public attitudes toward business had become increasingly positive by the late 1970s, particularly compared with the anticorporate sentiment of the late 1960s and early 1970s.

68. Bruce Shulman, "Slouching toward the Supply Side: Jimmy Carter and the New American Political Economy," in *The Carter Presidency: Policy Choices in the Post-New Deal Era*, eds. Gary M. Fink and Hugh Davis Graham (Lawrence: University Press of Kansas, 1998), 51–71.

69. James Tozzi, oral history, March 26, 2009, Nixon Oral History Project, Richard Nixon Presidential Library and Museum, National Archives and Records Administration, http://nixon.archives.gov/advancedsearch/search .php.

70. Susan Rose-Ackerman, "Progressivism and the Chicago School," in *Rethinking the Progressive Agenda: The Reform of the American Regulatory State* (New York: Free Press, 1992), 14–28.

71. Planning Subcommittee to the Air Quality Committee, MCA, memorandum of meeting minutes January 7–8, 1969, document no. CMA 085724, Chemical Industry Archives, a project of the Environmental Working Group, www.chemicalindustryarchives.org (hereafter cited as Chemical Industry Archives). "In support of the Committee's position that the cost:benefit ratio is an essential element in the determination of ambient air quality standards appropriate for any given situation" (4).

72. Bruce Ackerman and Richard B. Stewart, "Reforming Environmental Law," in *Foundations of Environmental Law and Policy*, ed. Richard L. Revesz (New York: Oxford University Press, 1997), 150–57.

73. Rose-Ackerman, "Progressivism."

74. NAS/NRC, *Principles for Evaluating Chemicals in the Environment: A Report of the Committee for the Working Conference on Principles of Protocols for Evaluating Chemicals in the Environment* (Washington, DC: NAS, 1975), 34.

75. H.G. Haight et al., "Risk-Benefit-Cost Background Report: Issues, Terms, Methodologies, and Activities," developed for Risk-Benefit-Cost Scoping Study Steering Group, June 1980, 7, document no. CMA 062837–71, Chemical Industry Archives. The steering group contended that with the great economic benefits the petrochemical industry bestows upon society come "risks of toxicity to man and his environment. In response to society's perception of these risks, the United States Congress in the 1970s enacted laws" to regulate chemicals (3). "Risk/Benefit/Cost Analysis" defines a chemical as "safe if its risks are judged to be acceptable" (5).

76. Vogel, *Fluctuating Fortunes*, 150.

77. Ibid., 198.

78. Ralph Harding to Fred C. Sutro Jr., USS Chemicals, February 26, 1975 (marked confidential), Executive Correspondences, Ralph Harding, SPI papers. According to the letter, Frank Ikard, president of the American Petroleum Institute, and Jack Roche, president of the American Iron and Steel Institute, handpicked all the members of the group. Harding noted to Sutro that attendance at Business Roundtable meetings should be disclosed within the SPI as an "informal meeting with Frank Ikard and Jack Roche."

79. Haight et al., "Risk-Benefit-Cost Background Report."

80. Vogel, *Fluctuating Fortunes*, 148–49.

81. Marc Karnis Landy, Marc J. Roberts, and Stephen R. Thomas, *The Environmental Protection Agency: Asking the Wrong Questions* (New York: Oxford University Press, 1990), 31.

82. Excerpts from letter from assistant secretary of OSHA Eula Bingham, EPA administrator Douglas M. Costle, FDA commissioner Donald Kennedy, and Consumer Product Safety Commission chairman S. John Byington to President Jimmy Carter, in "Cooperative Steps to Reform Regulatory Process and Improve the Health of Workers," *Public Health Reports* 92 (1977): 582.

83. Donald Kennedy, phone interview by author, October 22, 2010.

84. Merrill and Taylor, "Saccharin."

85. John Wargo, *Our Children's Toxic Legacy: How Science and Law Fail to Protect Us from Pesticides* (New Haven: Yale University Press, 1996), 113.

86. "F.D.A. Challenged on Plastic Bottle," *New York Times*, April 22, 1976, 35.

87. *Monsanto Co. v. Kennedy*, 198 U.S. App. D.C. 214 (1979).

88. Ibid.

89. Robert C. Barnard, "Some Regulatory Definitions of Risk: Interaction of Scientific and Legal Principles," *Regulatory Toxicology and Pharmacology* 11 (1990): 201–11.

90. Troetel, *Three-Part Disharmony*, 222–73.

91. Proctor, *Cancer Wars*, 58. Proctor quotes Samuel Epstein in his allegations that the Calorie Control Council's work was part of industry-led propaganda meant to undermine evidence of chemical carcinogens' threats.

92. Merrill and Taylor, "Saccharin"; Troetel, *Three-Part Disharmony*, 222–73.

93. Ibid.

94. Mary Lee Dunn, Polly Hoppin, and Beth Rosenberg, "Eula Bingham—Experience Bares 'the Real World' and Smart Politics Saves Lives," *New Solutions* 19 (2009): 81–93.

95. Transcript, F. Ray Marshall, interview by James Sterling Young, May 4, 1988, 2, Miller Center for Public Affairs, Carter Presidency Project, Presidential Oral History Program, University of Virginia, Charlottesville, Virginia, http://millercenter.org/president/carter/oralhistory/ray-marshall.

96. Transcript, Eula Bingham, interview by Judson MacLaury, January 5, 1981, no. 128, 11–13, Department of Labor Library, Washington, DC.

97. Ibid.

98. Landy, Roberts, and Thomas, *Environmental Protection Agency*, 182.

99. Philip Shabecoff, "Regulation by the U.S.: Its Costs vs. Its Benefits," *New York Times*, June 14, 1978.

100. Ibid.

101. Ralph Harding to Roy Albert, Institute of Environmental Medicine, NYU Medical Center, May 2, 1978, Executive Correspondences, Ralph Harding, SPI papers.

102. Meeting minutes, MCA Executive Committee, May 10, 1977, document no. CMA 061932, 3, Chemical Industry Archives. In the meeting, members discussed the possibility of developing research grants to the NAS or the NIH, as well as funding an expert conference with a "prominent educational institution such as M.I.T.," in order to help push risk-benefit and cost-benefit methods.

103. For a detailed discussion of the Interagency Regulatory Liaison Group (IRLG), see Landy, Roberts, and Thomas, *Environmental Protection Agency*, 172–203.

104. IRLG, "Scientific Bases for Identification of Potential Carcinogens and Estimation of Risk: Interagency Regulatory Liaison Group, Work Group on Risk Assessment," *Annual Review of Public Health* 1 (1980): 345–93.

105. Ibid.

106. Eula Bingham, e-mail to author, August 24, 2011.

107. Landy, Roberts, and Thomas, *Environmental Protection Agency*, 191–95.

108. Transcript, Eula Bingham, interview by Judson MacLaury, January 5, 1981, no. 128, 21–22, Department of Labor Library, Washington, DC.

109. Ralph Harding to Public Affairs Committee, SPI, "SPI Response to OSHA," memorandum, December 1, 1977, Committee Papers, SPI papers.

110. Ralph Harding, "American Industrial Health Council (AIHC) Strategy," memorandum, December 9, 1977, executive correspondences, Ralph Harding, SPI papers.

111. Ronald Lang, executive director, AIHC, to designated contacts, Associations Committee, Steering Committee, "Briefing for OSHA Witnesses—June 20th," memorandum, May 26, 1978, Committee Papers, SPI papers.

112. Harding, "American Industrial Health Council (AIHC) Strategy."

113. Ibid.

114. Proctor, *Cancer Wars*, 64.

115. Ibid.

116. Vice President Walter Mondale quoted in Bill Richards, "Study Sees 20% of Cancer Cases as Work-Related," *Washington Post*, September 12, 1978, A1.

117. Samuel P. Hays and Barbara D. Hays, *Beauty, Health, and Permanence: Environmental Politics in the United States, 1955–1985*, Studies in Environment and History (Cambridge: Cambridge University Press, 1987), 187.

118. Richard Doll and Richard Peto, *The Causes of Cancer: Quantitative Estimates of Avoidable Risks of Cancer in the United States Today* (Oxford: Oxford University Press, 1981). For a discussion of *The Causes of Cancer*, see Proctor, *Cancer Wars*, 68–69.

119. Sarah Boseley, "Renowned Cancer Scientist Was Paid by Chemical Firm for 20 Years," *Guardian* (U.K.), December 8, 2006. According to papers from Doll's archive at the Wellcome Foundation reviewed by Boseley, Doll received money from the Chemical Manufacturers Association for a review of vinyl chloride that dismissed charges of its carcinogenicity outside liver cancer, as well as a review of Agent Orange that concluded there was no evidence of carcinogenicity.

120. That same year, Carter also fired Joseph Califano and Treasury secretary W. Michael Blumenthal in a shake-up of the administration's cabinet members leading up to the 1980 election. Edward Walsh, "Califano, Blumenthal Are Fired, Bell Quits," *Washington Post*, July 20, 1979, A1.

121. *Industrial Union Department, AFL-CIO v. American Petroleum Institute* (the Benzene Case), 448 U.S. 607 (1980); ruling quoted in I. L. Feitshans, "Law and Regulation of Benzene," *Environmental Health Perspectives* 82 (1989): 301.

122. Ibid.; Paul M. Bangser, "An Inherent Role for Cost-Benefit Analysis in Judicial Review of Agency Decisions: A New Perspective on OSHA Rulemaking," *Boston College Environmental Affairs Law Review* 10 (1982): 365–443.

123. *American Textile Manufacturers Institute, Inc. v. Donovan*, 452 U.S. 490 (1981).

124. Bangser, "Inherent Role."

125. William J. Curran and Leslie Boden, "Occupational Health Values in the Supreme Court: Cost-Benefit Analysis," *American Journal of Public Health* 71 (1981): 1264–65.

126. Louis J. Casarett, Mary O. Amdur, Curtis J. Klaassen, and John Doull, eds., *Casarett and Doull's Toxicology: The Basic Science of Poisons*, 5th ed. (New York: McGraw-Hill Health Professions Division, 1996), 9.

3. REGULATORY TOXICITY TESTING AND
ENVIRONMENTAL ESTROGENS

1. "In Memoriam: David P. Rall, 1926–1999," *Environmental Health Perspectives* 107 (1999): A538–39.

2. National Institute of Environmental Health Sciences [NIEHS], *Environmental Health* (Washington, DC: Department of Health, Education and Welfare, Public Health Service, 1969), 7.

3. Robert Proctor, *Cancer Wars: How Politics Shapes What We Know and Don't Know about Cancer* (New York: Basic Books, 1995), 45.

4. A. L. Herbst, H. Ulfelder, and D. C. Poskanzer, "Adenocarcinoma of the Vagina: Association of Maternal Stilbestrol Therapy with Tumor Appearance in Young Women," *New England Journal of Medicine* 284, no. 15 (1971): 878–81.

5. Elizabeth Siegel Watkins, *The Estrogen Elixir: A History of Hormone Replacement Therapy in America* (Baltimore: Johns Hopkins University Press, 2007), 93.

6. "Chemical Profile: Bisphenol A," *Chemical Marketing Reporter*, January 4, 1999, 41; National Toxicology Program [NTP], *Carcinogenesis Bioassay of Bisphenol A (CAS No. 80–0507) in F344 Rats and B6C3F1 Mice (Feed Study)*, NTP Technical Report (Washington, DC: U.S. Department of Health and Human Services, Public Health Service, and NIH, 1982), 1.

7. James B. Knaak and Lloyd J. Sullivan, "Metabolism of Bisphenol A in the Rat," *Toxicology and Applied Pharmacology* 8 (1966): 175–84.

8. Division of Food Contact Substance Notification Review, Chemical Group 1, to Associate Directorate for Science and Policy, Attention, Mitchell Cheeseman, memorandum, August 13, 2001, obtained by author's request. (Parts of the memo are blacked out to protect confidential information.)

9. U.S. Department of Health and Human Services, FDA, "Update on Bisphenol A for Use in Food Contact Applications," January 2010, www.fda .gov/NewsEvents/PublicHealthFocus/ucm064437.htm.

10. "Bisphenol A (4,4'-Isopropylidenediphenol; 2,2-bis(4-Hydroxyphenyl) propane)," *American Industrial Hygiene Association Journal* 28 (1967): 301–4.

11. Knaak and Sullivan, "Metabolism of Bisphenol A."

12. René LeFaux, *Practical Toxicology of Plastics*, trans. Scripta Technica Ltd., English ed. (Cleveland, OH: CRC Press, 1968), 331.

13. Bryan Bilyeu, Witold Brostow, and Kevin Menard, "Halogen Containing Epoxy Composition and Their Preparation," University of North Texas,

March 10, 2009. Patents referenced in this patent application include Switzerland patent no. CH 211,116, issued to De Trey, on November 18, 1940, and titled "Verfahren zur Herstellung eines Hartbaren Kunstharzes"; Great Britain patent no. GB 518057, issued to De Trey Frères S.A. on February 15, 1940, and titled "A Process for the Manufacture of Thermo-setting Synthetic Resins by the Condensation of Alkylene Oxides with Anhydrides of Polybasic Acids"; and German patent no. DRP 749,512 (1938).

14. Irving S. Bengelsdorf, "Of Atoms and Men: Epoxy Resins Offer a Story of Success through Failures," *Los Angeles Times*, May 23, 1967, A6; Jack R. Ryan, "Tougher Plastics Invade Tool Field," *New York Times*, January 16, 1955, F2; R.L. Bowen, "Use of Epoxy Resins in Restorative Materials," *Journal of Dental Research* 35 (1956): 360–69.

15. "New Resin Liner Offered," *New York Times*, September 4, 1951, 37. An exact date for the introduction of epoxy resins into food containers could not be obtained. Plastic resins first began replacing tin as protective linings in food cans during the Second World War in response to a tin shortage and hoarding of canned foods. The War Production Board during the Second World War considered that the shortened shelf life of canned food with plastic liners would create a disincentive to hoarding. William Enright, "Tinless Cans Use Plastic Coatings," *New York Times*, August 11, 1940, 49; "WPB May Put Plastics in Food Cans to Conserve Tinplate," *Washington Post*, October 15, 1942, 6. Hoarding was discouraged by the federal government to keep food prices low as part of Roosevelt's economic stabilization order.

16. "Steel-Pipe Process Resists Corrosion," *New York Times*, July 8, 1962, 112.

17. Elvira Greiner, Thomas Kaelin, and Goro Toki, "Bisphenol A," in *Chemical Economics Handbook* (Stanford, CA: Stanford Research Institute, 2004), 1–37.

18. William F. Christopher and Daniel W. Fox, *Polycarbonates*, Reinhold Plastics Applications Series (New York: Reinhold, 1962), 4–5.

19. "Mobay Chemical Plastic," *Wall Street Journal*, July 7, 1959, 6.

20. "G.E. Picks Indiana Plant Site," *Chicago Daily Tribune*, June 5, 1959, C7.

21. "PC Looks at Lenses, Sees Bright Prospect," *Modern Plastics* 47 (September 1970): 82–83; Alan Hall, "Polycarbonate Shoots for 150 Million by 1980," *Modern Plastics* 47 (October 1970): 76–78; "Plastiscope," *Modern Plastics* 51 (May 1974): 104; "Polycarbonate Probes for Bigger Markets with New Molding Grades," *Modern Plastics* 50 (September 1973): 16–18.

22. Nelson Antosh, "From Hard Hats to Baby Bottles, Tough Plastics Spurs Expansion at Area Plants," *Houston Chronicle*, July 5, 1987, 1.

23. Sidney Gross, "Plastics Age Arrives Early," *Modern Plastics* 57 (February 1980): 45.

24. "BPA Production in 1985 Expected to Exceed Record," *Chemical Marketing Reporter*, December 2, 1985, 14.

25. NTP, *Carcinogenesis Bioassay of Bisphenol A*, 1.

26. J. Huff, "Long-Term Chemical Carcinogenesis Bioassays Predict Human Cancer Hazards: Issues, Controversies, and Uncertainties," *Annals of the New York Academy of Sciences* 895 (1999): 56–79.

27. Gregory J. Hart, director, U.S. General Accounting Office [GAO], to the Hon. Henry Waxman, House of Reps., March 30, 1979, GAO, www.gao.gov/index.html (hereafter cited as GAO documents).

28. Harold Stewart, consultant to the National Cancer Institute [NCI], to Wilhelm Hueper, June 13, 1973, in National Library of Medicine Archives, History of Medicine, Wilhelm Hueper Papers 1920–1981, Box 3, Folder: Correspondence 1966–1974.

29. Gerald E. Markowitz and David Rosner, *Deceit and Denial: The Deadly Politics of Industrial Pollution* (Berkeley: University of California Press, 2002), 215.

30. Joann S. Lublin, "Safety Problem," *Wall Street Journal,* February 21, 1978, 1; "3 Ex-Officials of Major Laboratory Convicted of Falsifying Drug Tests," *New York Times,* October 22, 1983, 1.

31. Lublin, "Safety Problem," 1; "3 Ex-Officials," 1; Markowitz and Rosner, *Deceit and Denial,* 215.

32. Dexter Goldman, "Chemical Aspects of Compliance with Good Laboratory Practices," in *Good Laboratory Practices: An Agrochemical Perspective,* ed. Willa Y. Garner and Maureen S. Barge (Washington, DC: American Chemical Society, 1988), 13–23.

33. Gregory J. Hart, director, GAO, to the Hon. Henry Waxman, House of Reps., Enclosure III, "NCI Has Not Adequately Monitored Tractor-Jitco's Bioassay Responsibilities," March 30, 1979, 33, GAO documents.

34. Ibid., 33–36.

35. GAO, *National Toxicology Program: Efforts to Improve Oversight of Contractors Testing Chemicals,* report to House Subcommittee on Oversight and Investigations Committee on Energy and Commerce, June 28, 1985; Edward A. Densmore, deputy director, Human Resources Division, "Three GAO Reviews of Contract Administration by the National Cancer Institute," statement to the Senate Subcommittee on Labor and Human Resources, June 2, 1981, www.legistorm.com/score_gao/show/id/10003.html; GAO, report to Hon. Henry A. Waxman, House of Reps., March 30, 1979, GAO documents; NTP, *Carcinogenesis Bioassay.*

36. Huff, "Long-Term Chemical Carcinogenesis."

37. NTP, *Carcinogenesis Bioassay.*

38. Statement of David Rall, director of the NIEHS and the NTP, in *National Toxicology Program: Hearing before the Subcommittee on Investigations and Oversight of the Committee on Science and Technology, U.S. House of Representatives,* 97th Cong., 2nd sess., July 15, 1981, 50, GAO documents.

39. GAO, *National Toxicology Program: Efforts to Improve Oversight of Contractors Testing Chemicals,* report to House Subcommittee on Oversight and Investigations Committee on Energy and Commerce, June 28, 1985, GAO documents.

40. The affiliation of Dr. E. Gordon is included as a footnote to his name in NTP, *Carcinogenesis Bioassay*.

41. Ibid., ix. A brief discussion of the involvement of the independent Peer Review Panel can be found in an article by James Huff of the NIEHS, a member of the panel. J. Huff, "Carcinogenicity of Bisphenol A Revisited," *Toxicological Sciences* 70 (2002): 281–84.

42. NTP, *Carcinogenesis Bioassay*, 46.

43. The first list of carcinogens published by the International Agency for Research on Cancer (IARC) was developed by Wilhelm Hueper and Erik Conway and by the World Health Organization. L. Tomatis, "The IARC Monographs Program: Changing Attitudes towards Public Health," *International Journal of Occupational and Environmental Health* 8 (2002): 144–52. The categories of evidence as outlined in 1983 included the following: "clear, some, equivocal, no evidence and inadequate experiment." James Huff, "A Historical Perspective on the Classification Developed and Used for Chemical Carcinogens by the National Toxicology Program during 1983–1992," *Scandinavian Journal of Work, Environment and Health* 18, suppl. 1 (1992): 74–82.

44. George Parris, *Bisphenol A: Preliminary Information Review* (Washington, DC: Dynamac Corporation for TSCA Interagency Testing Committee, 1982), EPA Office of Prevention, Pesticides and Toxic Chemical Library, Washington, DC, report no. 68–01–5789; EPA 560/ITC/82–0145.

45. The mutagenicity of epoxy resins was first reported in *Nature* in 1978: M. Andersen et al., "Mutagenic Action of Aromatic Epoxy Resins," *Nature* 276, no. 5686 (1978): 391–92.

46. J. Bitman and H.C. Cecil, "Estrogenic Activity of DDT Analogs and Polychlorinated Biphenyls," *Journal of Agricultural and Food Chemistry* 18 (1970): 1108–12; B.D. Hardin et al., "Testing of Selected Workplace Chemicals for Teratogenic Potential," *Scandinavian Journal of Work, Environment and Health* 7, suppl. 4 (1981): 66–75.

47. Parris, *Bisphenol A*, 17. Parris is currently the director of Environmental and Regulatory Affairs of the American Wood Preservers Institute.

48. J. Bitman et al., "Estrogenic Activity of o,p'-DDT in the Mammalian Uterus and Avian Oviduct," *Science* 162 (1968): 371–72; Bitman and Cecil, "Estrogenic Activity of DDT Analogs."

49. U.V. Solmssen, "Synthetic Estrogens and the Relation between Their Structure and Their Activity," *Chemical Review* 37 (1945): 481; H. Burlington and V.F. Lindeman, "Effect of DDT on Testes and Secondary Sex Characters of White Leghorn Cockerels," *Proceedings of the Society for Experimental Biology and Medicine* 74 (1950): 48–51.

50. Bitman and Cecil, "Estrogenic Activity of DDT Analogs," 1110.

51. Statement of David Rall, director of the NIEHS and the NTP, in *National Toxicology Program: Hearing*, July 15, 1981, 31.

52. "Kepone Mimics Female Hormone," *Science News*, January 19, 1980, 39.

53. F. M. Sullivan and S. M. Barlow, "Congenital Malformations and Other Reproductive Hazards from Environmental Chemicals," *Proceedings of the Royal Society of London B* 205 (1979): 91–110.

54. "PCBs Linked to Male Sterility," *Science News*, September 15, 1979, 183.

55. Hardin et al., "Testing of Selected Workplace Chemicals," 66.

56. Parris, *Bisphenol A*.

57. The Sprague-Dawley rat is a single genetic line of rats bred from the 1950s to the 1990s (approximately eight generations). F. S. vom Saal and C. Hughes, "An Extensive New Literature Concerning Low Dose Effects of Bisphenol A Shows the Need for a New Risk Assessment," *Environmental Health Perspectives* 113 (2005): 926–33.

58. Hardin et al., "Testing of Selected Workplace Chemicals," 69.

59. Bryan Hardin, Veritox Corporation, phone conversation with the author, June 22, 2006.

60. Ibid.

61. NTP, *Bisphenol A: Reproduction and Fertility Assessment in Cd-1 Mice When Administered in the Feed* (Research Triangle Park, NC: NTP, 1985), 14.

62. Ibid., 27.

63. Richard E. Morrissey et al., "The Developmental Toxicity of Bisphenol A in Rats and Mice," *Fundamental and Applied Toxicology* 8 (1987): 571–82.

64. Integrated Risk Information System (IRIS), EPA, "Bisphenol A. (CASRN 80–05–7)," September 26, 1988, revised July 1, 1993, www.epa.gov/iris/subst/0356.htm.

65. The 50 mg/kg/day dose was based on an estimate of the daily food consumed that was contaminated with 1,000 ppm bisphenol A. (Ibid.)

66. Ibid.

67. R. R. Newbold and J. A. McLachlan, "Vaginal Adenosis and Adenocarcinoma in Mice Exposed Prenatally or Neonatally to Diethylstilbestrol," *Cancer Research* 42 (1982): 2003–11.

68. John McLachlan, speech at the Gordon Research Conference on Environmental Endocrine Disruptors, June 19, 2008, which the author attended; John A. McLachlan, "Environmental Signaling: What Embryos and Evolution Teach Us about Endocrine Disrupting Chemicals," *Endocrine Reviews* 22 (2001): 319–34.

69. J. A. McLachlan, R. R. Newbold, and B. Bullock, "Reproductive Tract Lesions in Male Mice Exposed Prenatally to Diethylstilbestrol," *Science* 190 (1975): 991–92.

70. J. A. McLachlan, "Prenatal Exposure to Diethylstilbestrol in Mice: Toxicological Studies," *Journal of Toxicology and Environmental Health* 2 (1977): 527–37; J. A. McLachlan, M. Metzler, and J. C. Lamb, "Possible Role of Peroxidase in the Diethylstilbestrol-Induced Lesions of the Syrian Hamster Kidney," *Life Sciences* 23 (1978): 2521–24; J. C. Lamb, R. R. Newbold, and J. A. McLachlan, "Evaluation of the Transplacental Toxicity of Diethylstil-

bestrol with the Scanning Electron Microscope," *Journal of Toxicology and Environmental Health* 5 (1979): 599–603; J. A. McLachlan, R. R. Newbold, and B. C. Bullock, "Long-Term Effects on the Female Mouse Genital Tract Associated with Prenatal Exposure to Diethylstilbestrol," *Cancer Research* 40 (1980): 3988–99.

71. A. L. Herbst, "Summary of the Changes in the Human Female Genital Tract as a Consequence of Maternal Diethylstilbestrol Therapy," *Journal of Toxicology and Environmental Health*, suppl. 1 (1976): 13–20; A. L. Herbst, R. E. Scully, and S. J. Robboy, "Vaginal Adenosis and Other Diethylstilbestrol-Related Abnormalities," *Clinical Obstetrics and Gynecology* 18 (1975): 185–94.

72. Proctor, *Cancer Wars*, 225–26.

73. A. M. Soto and C. Sonnenschein, "The Somatic Mutation Theory of Cancer: Growing Problems with the Paradigm?" *Bioessays* 26 (2004): 1097–107.

74. Fujimura's "theory method package" reveals how very controlled and standardized mouse strains provided support to the oncogene theory itself. As a result, any "noise" in the data can be quieted. Joan Fujimura, *Crafting Science: A Sociohistory of the Quest for the Genetics of Cancer* (Cambridge, MA: Harvard University Press, 1996). As an example for understanding how this gene-based comprehension of cancer and development shaped the conceptualization of environment, consider wild moles, animals not used in laboratory experiments. Moles born in the summer have brown coats and those born in the winter have white coats. The environmental influence on the phenotype of the mole disrupts the theory method package provided by standardized mouse strains. "Such animals without any such seasonality provided better models for studying stability in the relationship between DNA and phenotype." Ana Soto, Tufts University, Sackler School of Graduate Biomedical Sciences, Boston, interview by author, April 17, 2007.

75. Ames at first supported the notion that a single molecule of a carcinogen could cause cancer, the so-called single-hit theory that lent legitimacy and theoretical support to the Delaney clause. But by the late 1970s, Ames became an outspoken critic of environmental efforts to regulate industrial chemicals. As Proctor details, in the early 1980s Ames developed the natural carcinogen thesis, which argues that cancer-causing substances exist in the natural world and that the environmental movement's focus on synthetic substances overlooks this problem. David Proctor, "Natural Carcinogens and the Myth of Toxic Hazards," in *Cancer Wars*; Jerry E. Bishop, "Cancer Dilemma: New Laboratory Tests for Chemicals' Safety Stir Scientific Debate," *Wall Street Journal*, May 16, 1977, 1.

76. Proctor, *Cancer Wars*, 150.

77. Senate Committee on Labor and Public Welfare, *National Program for the Conquest of Cancer: Report of the National Panel of Consultants on the Conquest of Cancer*, authorized by S. Res. 376, agreed to by Senate April 27, 1970, 120. Italics in original.

78. Ibid., 73.

79. Ibid., 78.

80. Marla Cone, "President's Cancer Panel: Environmentally Caused Cancers Are 'Grossly Underestimated' and 'Needlessly Devastate American Lives,'" *Environmental Health News,* May 6, 2010, www.environmental healthnews.org/ehs/news/presidents-cancer-panel; Suzanne H. Ruben for the President's Cancer Panel, *Reducing Environmental Cancer Risk: What We Can Do Now, 2008–09 Annual Report,* U.S. Department of Health and Human Services, NIH, NCI.

81. C. Sonnenschein and A. M. Soto, *The Society of Cells: Cancer Control of Cell Proliferation* (Oxford: Bios Scientific Publisher, 1999), 63.

82. For a detailed discussion, see ibid.

83. Ibid.

84. Nelly Oudshoorn, "The Measuring of Sex Hormones," in *Beyond the Natural Body* (New York: Routledge, 1994).

85. Sonnenschein and Soto, *Society of Cells,* 60–77.

86. Ana Soto, Tufts University, Sackler School of Graduate Biomedical Sciences, Boston, interview by author, April 17, 2007.

87. Ibid.

88. McLachlan recalled this early research period as a "lonely" time. Sheldon Krimsky, *Hormonal Chaos: The Scientific and Social Origins of the Environmental Endocrine Hypothesis* (Baltimore: Johns Hopkins University Press, 2000, 13. Ludwig Fleck defined a "thought collective" in 1935 "as *a community of persons mutually exchanging ideas or maintaining intellectual interaction . . . It also provides the specific 'carrier' for the historical development of any field of thought, as well as for the given stock of knowledge and level of culture.*" Italics in original. Ludwig Fleck, *Genesis and Development of a Scientific Fact* (Basel: Benno Schwabe, 1935), 39.

89. John McLachlan and David Rall, "Potential for Exposure to Estrogens in the Environment," in *Estrogens in the Environment: Proceedings of the Symposium on Estrogens in the Environment, Raleigh, NC, September 10–12, 1979,* ed. John McLachlan (Amsterdam: Elsevier/North Holland, 1980), 199.

90. McLachlan, "Environmental Signaling."

91. Jack Gorski, "Models of Estrogenic-Hormone Activity," in McLachlan, *Estrogens in the Environment,* 3–9.

92. J. A. McLachlan, ed., *Estrogens in the Environment II: Influences on Development,* Laboratory of Reproductive and Developmental Toxicology, NIEHS, April 10–12, 1985 (New York: Elsevier, 1985).

93. J. A. McLachlan and K. S. Korach, "Symposium on Estrogens in the Environment, III," *Environmental Health Perspectives* 103, suppl. 7 (1995): 3–4.

94. George Lardner Jr. and Joby Warrick, "Pesticide Coalition's Text Enters House Bill; Industry, Farmers Trying to Blunt U.S. Regulations," *Washington Post,* May 13, 2000, A1; Dan Fagin, Marianne Lavelle, and Center for Public Integrity, *Toxic Deception: How the Chemical Industry Manipulates Science,*

Bends the Law, and Endangers Your Health (Secaucus, NJ: Carol Publishing Group, 1996), 117.

95. "The Weinberg Group, Inc.," InsideView, 2012, www.insideview.com/directory/the-weinberg-group-inc.

4. ENDOCRINE DISRUPTION

1. "Chemical Profile: Bisphenol A," *Chemical Marketing Report* 244 (1993): 45.

2. Aruna V. Krishnan et al., "Bisphenol-A: An Estrogenic Substance Is Released from Polycarbonate Flasks during Autoclaving," *Endocrinology* 132 (1993): 2279–86. General Electric produced a BPA-based polycarbonate plastic used to make laboratory flasks. When the Stanford researchers queried General Electric about the problem, the company informed them that it was aware of BPA's estrogenicity but provided the researchers with a bottle-washing technique designed to reduce the migration of the compound and repeated that the low levels released were safe.

3. A.M. Soto, H. Justica, J.W. Wray, and C. Sonnenschein, "p-Nonyl-phenol: An Estrogenic Xenobiotic Released from 'Modified' Polystyrene," *Environmental Health Perspectives* 92 (May 1991): 167–73. Whereas GE acknowledged the presence and estrogenicity of BPA, the Tufts researchers were not told what chemical might be causing the estrogenic effects. The researchers themselves were then required to conduct chemical detective work, teaming up with analytical chemists to identify different compounds and then testing their estrogenicity. The researchers' investigation is detailed in Theo Colborn, Dianne Dumanoski, and John Peterson Myers, *Our Stolen Future: Are We Threatening Our Fertility, Intelligence, and Survival? A Scientific Detective Story* (New York: Penguin Books, 1996), 122–30.

4. Jon Cohen, "Lab Accident Reveals Potential Health Risk of Common Compound," *Science* 300 (April 4, 2003): 31–32.

5. P.A. Hunt et al., "Bisphenol A Exposure Causes Meiotic Aneuploidy in the Female Mouse," *Current Biology* 13 (2003): 546–53.

6. Kenneth S. Korach, "Editorial: Surprising Places of Estrogenic Activity," *Endocrinology* 132 (1993): 2277–78.

7. Theo Colborn, interview by author, Paonia, CO, September 23, 2006. In her dissertation research, Colborn compared concentrations of heavy metals in the exoskeleton of invertebrates living in area streams with heavy metals concentrations in the water. She developed a model for using invertebrates as indicators of changing concentrations of metals in water. Toward the end of Colborn's graduate work, Richard Lamm, governor of Colorado, appointed her to the state's Natural Areas Council.

8. Ibid.

9. Michael Gilbertson, "Water Quality Objectives: Yardsticks of the Great Lakes Water Quality Agreement," *Environmental Health Perspectives* 107 (1999): 239–41; Michael Gilbertson, "Scientific Issues in Relation to

Lakewide Management Plan: Linking Science and Policy," *Environmental Health Perspectives* 108 (2000): 467–68; Joseph L. Jacobson and Sandra W. Jacobson, "Effects of Exposure to PCBs and Related Compounds on Growth and Activity in Children," *Neurotoxicology and Teratology* 12 (1990): 319–26; Joseph L. Jacobson, Sandra W. Jacobson, and Harold E. B. Humphrey, "Effects of in Utero Exposure to Polychlorinated Biphenyls and Related Contaminants on Cognitive Functioning in Young Children," *Journal of Pediatrics* 116 (1990): 38–45; Sandra W. Jacobson et al., "The Effect of Intrauterine PCB Exposure on Visual Recognition Memory," *Child Development* 56 (1985): 853–60.

10. EPA, "William Reilly: Biography," press release, February 2, 1989, www.epa.gov/aboutepa/history/admin/agency/reilly.html.

11. Samuel P. Hays and Barbara D. Hays, *Beauty, Health, and Permanence: Environmental Politics in the United States, 1955–1985*, Studies in Environment and History (Cambridge: Cambridge University Press, 1987), 276, 409, 419–22.

12. Resources for the Future, "RFF Honors Terry Davies for Influential Role in Environmental Policymaking," *Resources*, no. 140 (Summer 2000): 18–19, http://rff.org/rff/documents/rff-resources-140-davies.pdf. From 1991 to 1992, Davies was the executive director of the National Commission on the Environment, which outlined the future of U.S. environmental policy.

13. Richard N. L. Andrews, *Managing the Environment, Managing Ourselves: A History of American Environmental Policy* (New Haven: Yale University Press, 1999).

14. Theo Colborn, interview by author, Paonia, CO, September 23, 2006.

15. M. Gilbertson, T. Kubiak, J. Ludwig, and G. Fox, "Great Lakes Embryo Mortality, Edema, and Deformity Syndrome (GLEMEDS) in Colonial Fish Eating Birds: Similarity to Chick-Edema Disease," *Journal of Toxicology and Environmental Health* 33 (1991): 455–520.

16. Theo Colborn, interview by author, Paonia, CO, September 23, 2006.

17. A. B. Hill, "The Environment and Disease: Association or Causation?" *Proceedings of the Royal Society of Medicine* 58 (1965): 297.

18. John Frank to Working Group of the Science Advisory Board of the International Joint Commission, " 'Weight of Evidence' Approach to Assessing the Adverse Effects of Environmental Hazards," memorandum, February 24, 1994, in Theo Colborn, personal papers, Paonia, CO, uncatalogued collection (hereafter cited as TC papers). Colborn used this method of weight-of-the-evidence assessment in her own work on the IJC environmental health assessment report.

19. Hill, "Environment and Disease," 299–300.

20. Ibid. The chi-squared is a statistical test used to compare observed data with the data expected on the basis of a given thesis, and using categorical data (e.g., "yes" or "no").

21. C. F. Cranor, "Learning from the Law to Address Uncertainty in the Precautionary Principle," *Science and Engineering Ethics* 7 (2001): 313–26. The statistical power of a study is determined mathematically as 1—ß (ß

= type II error). A type II error is the probability of accepting the null (no effect, no relationship between variables, or negative result in a screen) when in fact there is a relationship, disease, or effect: a false negative. The power is the probability of rejecting the null hypothesis (that there is no relationship) when the null is false. Put another way, it is the probability of not committing type II errors. A type I error is the probability of rejecting the null when in fact no relationship exists: a false positive. The sample size, power, type I (alpha, false positive), and type II (false negative) are all mathematically related. There is an inherent moral decision in privileging studies with stronger statistical correlations because in exchange for a higher relative risk, for example, holding sample size and type I error constant, there is a higher chance of committing a type II error. In the case of toxic chemicals, a type II error translates to saying there is no relationship between a chemical and an effect when in fact one exists.

22. Rachel Carson, *Silent Spring* (1962; repr., Boston: Houghton Mifflin, 1994), 189.

23. Rosalie Bertell, "Weight of Evidence versus Proof of Causation," in *Applying Weight of Evidence: Issues and Practice*, ed. Michael Gilbertson and Sally Cole-Misch (Windsor, ON: International Joint Commission, 1994).

24. D. Gee, "Late Lessons from Early Warnings: Toward Realism and Precaution with Endocrine-Disrupting Substances," *Environmental Health Perspectives* 114, suppl. 1 (2006): 152–60; Timothy O'Riordan and James R. Cameron, *Interpreting the Precautionary Principle* (London: Earthscan, 1994).

25. United Nations Environment Programme, Rio Declaration on Environment and Development, 1992, www.unep.org/Documents.Multilingual/Default.asp?documentid=78&articleid=1163.

26. Theo Colborn, *Great Lakes, Great Legacy?* (Washington, DC: Conservation Foundation, 1990).

27. Theo Colborn, interview by author, Paonia, CO, September 23, 2006.

28. Ibid.

29. From 1997 to 2000 I worked under Pete Myers for the W. Alton Jones Foundation, where I first learned about endocrine disruption. In 2002 I worked briefly with Colborn at the World Wildlife Fund before she left to start her own organization.

30. Sheldon Krimsky, *Hormonal Chaos: The Scientific and Social Origins of the Environmental Endocrine Hypothesis* (Baltimore: Johns Hopkins University Press, 2000), 23–24. Early one morning at a board meeting of the Jones Foundation in the early 1990s, Colborn sat down with then secretary of the Interior Babbitt, who had been asked to give the keynote speech and who was familiar with Colborn's concerns about endocrine-disrupting chemicals. Colborn recalls Babbitt quietly turning to her and asking, "Remind me, Theo, what is the endocrine system?" Theo Colborn, interview by author, Paonia, CO, September 23, 2006.

31. Arthur L. Herbst and Howard Alan Bern, *Developmental Effects of Diethylstilbestrol (DES) in Pregnancy* (New York: Thieme-Stratton, 1981).

32. Theo Colborn, interview by author, Paonia, CO, September 23, 2006.

33. Intergovernmental Panel on Climate Change, "First Assessment Report: Overview," 1990, www.ipcc.ch/ipccreports/1992%20IPCC%20Supplement/IPCC_1990_and_1992_Assessments/English/ipcc_90_92_assessments_far_overview.pdf.

34. N. Oreskes and Erik M. Conway, "Defeating the Merchants of Doubt," *Nature* 465 (2010): 686–87.

35. John Peterson Myers, personal communication to author, Bisphenol A Meeting, Center for the Evaluation of Risks to Human Reproduction, Hilton Hotel, Old Town Alexandria, Virginia, August 7, 2007.

36. The full consensus statement is published in Theo Colborn and Coralie Clement, eds., *Chemically-Induced Alterations in Sexual and Functional Development: The Wildlife/Human Connection*, Advances in Modern Environmental Toxicology 21 (Princeton, NJ: Princeton Scientific Publishing, 1992), 1–8.

37. Various definitions of endocrine disruption exist, and the differences are an important part of the debate about the risks of these compounds. Krimsky, *Hormonal Chaos*; Sheldon Krimsky, "An Epistemological Inquiry into the Endocrine Disruptor Thesis," *Annals of the New York Academy of Sciences* 948 (2001): 130–42; A. C. Gore, J. J. Heindel, and R. T. Zoeller, "Endocrine Disruption for Endocrinologists (and Others)," *Endocrinology* 147, no. 6, suppl. 1 (2006): S1–3.

38. Susan P. Porterfield, *Endocrine Physiology* (St. Louis, MO: Mosby, 1997).

39. Theo Colborn, interview by author, Paonia, CO, September 23, 2006.

40. Howard Bern, "The Fragile Fetus," in Colborn and Clement, *Chemically-Induced Alterations*, 9–15.

41. Theo Colborn, Frederick S. vom Saal, and Ana M. Soto, "Developmental Effects of Endocrine-Disrupting Chemicals in Wildlife and Humans," *Environmental Health Perspectives* 101 (1993): 378–84.

42. Kimberly G. Thigpen, "Spheres of Influence: Presidential Scorecard," *Environmental Health Perspectives* 102 (1994): 370–71.

43. Gary Lee, "In Food Safety Changes, Victories for Many," *Washington Post*, July 28, 1996, A4.

44. Chuck Benbrook, personal communication to author, July 20, 2007; John Wargo, *Our Children's Toxic Legacy: How Science and Law Fail to Protect Us from Pesticides* (New Haven, CT: Yale University Press, 1996), 70–78, 89–90.

45. Wargo, *Our Children's Toxic Legacy*, 120–23.

46. Krimsky, *Hormonal Chaos*, 205.

47. Deborah Cadbury, *The Estrogenic Effect: Assault on the Male* (London: BBC, 1993), DVD, available from Films Media Group, http://ffh.films.com/id/9504/The_Estrogen_Effect_Assault_on_the_Male.htm.

48. Krimsky, *Hormonal Chaos*, 68–74.

49. Theo Colborn, interview by author, Paonia, CO, September 23, 2006.

50. Provisions for a screening program for estrogenic chemicals are included in the Food Quality Protection Act of 1996 (FQPA), 7 USC 136, sec. 304, and the Safe Drinking Water Act (SDWA) Amendments of 1996, P.L. 104–182, sec. 136. Krimsky, *Hormonal Chaos*, 68–72. The transformation of the theory of endocrine disruption from a marginal scientific thesis into what Krimsky calls a "public hypothesis" occurred in the first half of the 1990s through a combination of congressional hearings, popular media coverage, the increased involvement and support of the Clinton administration, the support of powerful breast cancer and environmental organizations, and the passage of key legislation in 1996 that first introduced the need for the regulation of estrogenic chemical compounds.

51. Frederick L. Webber, president and CEO, Chemical Manufacturers Association, to Larry Thomas, Society of the Plastics Industry [SPI]; Jay Vroom, American Crop Protection Association; Red Cavaney, American Plastics Council; and Kip Howlett, Chlorine Chemistry Council, "Endocrine Issues Coalition Steering Committee," memorandum, October 13, 1995, TC papers.

52. Dow Chemical Company, "Position on Endocrine Disruptors" (included in packet from Frank Popoff, chairman and CEO, Dow Chemical Company), February 15, 1994, TC papers.

53. Jerome Heckman, Heckman and Keller, LLC, interview by author, July 15, 2010.

54. "Scientific Consensus on Endocrine Disruption," 2005, http://ourstolenfuture.org/Consensus/consensus.htm. The conference's additional consensus statements on endocrine disruption include Wingspread 1993, Wingspread 1995—I, Wingspread 1995—II, Erice 1995, Yokohama 1999, Prague Declaration 2005, and Vallombrosa 2005.

55. The FQPA and the SDWA were signed into law on August 6, 1996. *Our Stolen Future* was released in the spring of 1996. In a confidential e-mail to Theo Colborn, an individual working at a major pesticide manufacturer alleged that Vice President Gore's office had struck an agreement with the industry trade associations: if they responded well to the publication of *Our Stolen Future*, they would get their "Delaney fix." Confidential e-mail to Colborn, author and date blacked out, TC papers.

56. Tina Adler, "The Clinton-Gore Agenda: Environmental Pledges and Economic Puzzles," *Environmental Health Perspectives* 101 (1993): 30–31.

57. Exposures of up to 1 ppb require additional genetic mutation studies, and above that level chronic carcinogenicity and reproductive toxicity studies must be undertaken. All of the regulatory exemptions remain in place, including migration below 0.5 ppb; the Threshold of Regulation rule; prior sanctioned; not reasonably expected to migrate into food (dependent on the detection limit); and generally recognized as safe (GRAS) compounds. If a substance does not fall under any of these exemptions, the manufacturer must submit a food contact notification. Jerome Heckman and Deborah Ziffer, "Fathoming Food Packing Regulations Revisited," *Food and Drug Law Journal* 56 (2001): 179–95.

58. Jerome Heckman, Heckman and Keller, LLC, interview by author, July 15, 2010.

59. Jerome Heckman, "Food Contact Substances and the Need for Food Safety Reform," *Regulatory Toxicology and Pharmacology* 24 (1996): 236–39.

60. Lynn Goldman, Johns Hopkins University, phone interview by author, August 15, 2007. D'Amato's original language called for the testing of all chemicals for estrogenicity with a single testing system. The EPA included multiple testing systems and other scientific information to broaden the scope of the regulatory process.

61. EPA, Endocrine Disruptor Testing and Screening Advisory Committee [EDSTAC], "Executive Summary," in *Endocrine Disruptor Testing*, 1.

62. Ibid.

63. Confidential e-mails to Theo Colborn, February 6 and 9, 1996, author and date blacked out, TC papers. "Alar scare" refers to an industry-led effort to block the banning of the pesticide Alar, used most notoriously on apples, in the mid-1980s. Dan Fagin, Marianne Lavelle, and Center for Public Integrity, *Toxic Deception: How the Chemical Industry Manipulates Science, Bends the Law, and Endangers Your Health* (Secaucus, NJ: Carol Publishing Group, 1996), 107–8.

64. Jon Holtzman, vice president of communications, Chemical Manufacturers Association, memorandum, March 6, 1996, TC papers.

65. D. Michaels, "Doubt Is Their Product," *Scientific American* 292, no. 6 (2005): 96–101.

66. David Michaels, *Doubt Is Their Product: How Industry's Assault on Science Threatens Your Health* (Oxford: Oxford University Press, 2008); Naomi Oreskes and Erik M. Conway, *Merchants of Doubt: How a Handful of Scientists Obscured the Truth on Issues from Tobacco Smoke to Global Warming* (New York: Bloomsbury Press, 2010).

67. Richard N. Cooper, "Book Reviews: *Risky Business: An Insider's Account of the Disaster at Lloyd's of London* by Elizabeth Luessenhop and Martin Mayer; *Ultimate Risk: The Inside Story of the Lloyd's Catastrophe* by Adam Raphael," *Foreign Affairs* 75 (1996): 136–67.

68. Peter Huber, *Galileo's Revenge: Junk Science in the Courtroom* (New York: Basic Books, 1991).

69. Ralph Harding, "Product Liability Strategy Group: A Platform and a Program for Product Liability Solutions," draft memorandum, January 31, 1977, executive officers' correspondences, and Ralph Harding to Secretary of Commerce Elliot L. Richardson, June 9, 1976, executive officers' correspondences, both in Hagley Museum and Archives, Soda House, Society of the Plastics Industry [SPI], Inc., Records 1936–1987, Committee Records (hereafter referred to as SPI papers). The letter was also carbon-copied to William Ruckelshaus, former head of the EPA, then working as a private lawyer.

70. Herbert L. Needleman, "The Removal of Lead from Gasoline: Historical and Personal Reflections," *Environmental Research* 84 (2000): 20–35.

71. M. A. Berger, "What Has a Decade of Daubert Wrought?" *American Journal of Public Health* 95, suppl. 1 (2005): S59–65; S. Jasanoff, "Law's Knowledge: Science for Justice in Legal Settings," *American Journal of Public Health* 95, suppl. 1 (2005): S49–58; T. O. McGarity, "Daubert and the Proper Role for the Courts in Health, Safety, and Environmental Regulation," *American Journal of Public Health* 95, suppl. 1 (2005): S92–98.

72. Roni A. Neff and Lynn R. Goldman, "Regulatory Parallels to *Daubert:* Stakeholder Influence, 'Sound Science,' and the Delayed Adoption of Health-Protective Standards," *American Journal of Public Health* 95, suppl. 1 (2005): S81–91.

73. A. Baba et al., "Legislating 'Sound Science': The Role of the Tobacco Industry," *American Journal of Public Health* 95, suppl. 1 (2005): S20–27; Chris Mooney, "Some Like It Hot," *Mother Jones* (May–June 2005): 36–94; Neff and Goldman, "Regulatory Parallels to *Daubert*"; Michaels, *Doubt Is Their Product;* Oreskes and Conway, *Merchants of Doubt.*

74. Michaels, *Doubt Is Their Product,* 58.

75. APCO Associates to TASSC [The Advancement of Sound Science Coalition] Supporters, "TASSC Activities," memorandum, April 28, 1995, document no. 2048294346, Phillip Morris USA Document Site, www.pmdocs .com. Michaels writes that TASSC officials "reached out to executives at Proctor and Gamble, General Motors, 3M, Dow Chemical, and other corporations." Michaels, *Doubt Is Their Product,* 85.

76. Mooney, "Some Like It Hot"; Paul D. Thacker, "Smoked Out," *New Republic,* February 6, 2006, 13–14.

77. Steven Milloy, "Coming Soon: More Chemical Scares Than Anyone Dreamed Possible," FoxNews.com, May 20, 2001.

78. "Science Watchdog Group Celebrates Third Anniversary with Renewed Commitment to Exposing the Use of Junk Science," *PR Newswire,* December 3, 1996.

79. Elizabeth Whelan, research associate, Harvard School of Public Health, and executive director, American Council on Science and Health, to Ralph Harding, president, SPI, October 10, 1977, executive officers' correspondences, SPI papers.

80. Gerald E. Markowitz and David Rosner, *Deceit and Denial: The Deadly Politics of Industrial Pollution* (Berkeley: University of California Press, 2002), 288–89.

81. Jerome Heckman, Keller and Heckman, to Elizabeth Whelan, research associate, Harvard School of Public Health, and executive director, American Council on Science and Health, October 6, 1977, and Ralph Harding to Dorothy Parker, Atheneum Publishers, July 21, 1976, both in executive officers' correspondences, SPI papers.

82. Elizabeth Whelan, William London, and Leonard Flynn, "ACSH Commentary on 'Our Stolen Future,'" *American Council on Science and Health,* May 31, 2006, www.acsh.org/healthissues/newsID.1346/healthissue_detail .asp.

83. S.F. Arnold et al., "Synergistic Activation of Estrogen Receptor with Combinations of Environmental Chemicals," *Science* 272 (1996): 1489–92; J.A. McLachlan, "Synergistic Effect of Environmental Estrogens: Report Withdrawn," *Science* 277 (1997): 462–63.

84. On numerous occasions in debates on endocrine disruption or even BPA, this incident was raised as reason for doubt and skepticism about the entire research field.

85. Anne N. Hirschfield, Michael F. Hirschfield, and Jodi A. Flaws, "Problems beyond Pesticides," *Science* 272 (1996): 1445.

86. Gina Kolata, "Chemicals That Mimic Hormones Spark Alarm and Debate," *New York Times*, March 19, 1996, C1, C10. During a private meeting with several well-established reporters at the *New York Times*, including science writer Nicholas Wade, the book's authors learned just how deep the skepticism ran. Wade declared that the book didn't present "real science." Yet according to the authors' recollections of the meeting, Wade, when asked, admitted that he had yet to actually read the book. Pete Myers, personal communication to author, June 10, 2008; Dianne Dumanoski, e-mail to author, June 24, 2008.

87. Krimsky, *Hormonal Chaos*, 209. Federally chartered committees are required by law to hold open hearings, engage in stakeholder participation, and notify the public about committee meetings.

88. EPA/EDSTAC, introduction to *Endocrine Disruptor Testing*, 2.

89. The original mandate for EPA under the FQPA and SDWA was to develop a screening and testing program for environmental estrogens. The administrator expanded this to include interaction with androgens and thyroid hormones. Krimsky, *Hormonal Chaos*, 211. As Krimsky details, the committee was faced with not only the eighty-seven thousand chemicals on the market but possible chemical mixtures as well. Testing even a fraction of these chemicals and combinations of chemicals would take thousands of years.

90. National Research Council (NRC), Committee on Hormonally Active Agents in the Environment, *Hormonally Active Agents in the Environment* (Washington, DC: National Academy Press, 1999). The introduction to the report discusses the "issues that divided the committee," including the definition of an endocrine disruptor. The report concludes that such divisions represent "epistemiologic" differences among members (15).

91. EPA/EDSTAC, Appendix A, in *Endocrine Disruptor Testing*.

92. Christopher Borgert, "Draft Definition," February 2, 1998, e-mail to EDSTAC listserv, TC papers.

93. Chemical Manufacturers Association to Public Information and Records Integrity Branch, Information Resources and Services Division, Office of Pesticide Programs, EPA, "Comments to the Science Advisory Panel and the Science Advisory Board on the April 3, 1998, Draft Report of the Endocrine Disruptor Screening and Testing Advisory Committee," April 27, 1998, TC papers (last page missing).

94. Hugh Patrick Toner, vice president of technical affairs, SPI, to Public Information and Records Integrity Branch, Information Resources and Services Division, Office of Pesticide Programs, EPA, April 27, 1998, TC papers.

95. Minority Report from Frederick vom Saal to National Research Committee/National Academy of Science (NRC/NAS) Hormonally Active Agents Committee, "Critique of the Majority Executive Committee," August 14, 1998, Minority Report from Ana M. Soto to NRC/NAS Hormonally Active Agents Committee, August 14, 1998, and Louis Guillette to Ernst Knobil, chair, NRC/NAS Hormonally Active Agents Committee, "RE: Report on HAAs," August 14, 1998, all in TC papers.

96. Goldman quoted in Krimsky, *Hormonal Chaos*, vii.

97. Lynn Goldman, Johns Hopkins University, phone interview by author, August 15, 2007.

98. Lynn Goldman to EDSTAC members, "Endocrine Disruptor—The Definition Revisited," memorandum, March 2, 1998, TC papers.

99. Don Lamb, Bayer Corporation; Jim Quance, Exxon Chemical Corporation; Joe LeBeau, consultant representing the Chlorine Chemistry Council; Angelina Duggan, FMC Corporation; Abe Tobia, BASF Corporation; George Daston, Procter & Gamble; Tom Osmitz, S. C. Johnson & Son; and Chris Borgert, consultant representing small business, to Lynn R. Goldman, assistant administrator, Office of Pesticide Programs, EPA, "Definition of Endocrine Disruption and Potential Endocrine Disruptor," memorandum, March 13, 1998, TC papers. The industry representatives' memo closes by stating that by including the term *potential endocrine disruptor* Goldman and the EPA "risk destroying consensus on the entire program. Therefore, we strongly urge you to re-consider your proposal."

100. NRC, *Hormonally Active Agents*, 309–10.

101. Lynn Goldman, Johns Hopkins University, phone interview by author, August 15, 2007.

102. EPA/EDSTAC, "Executive Summary," in *Endocrine Disruptor Testing*, 1.

103. Hugh Patrick Toner, vice president of technical affairs, SPI, to Public Information and Records Integrity Branch, Information Resources and Services Division, Office of Pesticide Programs, EPA, April 27, 1998, TC papers. In this public letter submitted to the EPA, the SPI "emphatically maintains" that a chemical should not be classified as an endocrine disruptor if it is found to test positively in the first tier of screens. The letter suggested that the precautionary principle promotes unsound science: "The principle may easily overwhelm and preempt the opportunity for sound scientific research . . . SPI and others have made substantial investment in a sound science approach to research in this field. SPI is confident that science and technology, given time, can substantially respond to health, safety, and environmental concerns now and in the future."

104. NRC, *Hormonally Active Agents*, 16.

105. Multiple definitions of endocrine disruptors are discussed in Richard Evans et al., "State of the Art Assessment of Endocrine Disrupters: 2nd Interim Report, Part 1: Summary of the State of the Science," contract no. 070307/2009/550687/SER/D3, European Commission, 2009, 8–10.

106. EPA/EDSTAC, "Screening and Testing," in *Endocrine Disruptor Testing*.

107. Thomas Kuhn, *The Structure of Scientific Revolutions* (Chicago: University of Chicago Press, 1962), 42.

5. THE LOW-DOSE DEBATE

1. F. S. vom Saal and F. Bronson, "Sexual Characteristics of Adult Female Mice Are Correlated with Their Blood Testosterone Levels during Prenatal Development," *Science* 208 (1980): 597–99; F. S. vom Saal, "The Intrauterine Position Phenomenon: Effects on Physiology, Aggressive Behavior, and Population Dynamics in House Mice," *Progress in Clinical Biology Research* 169 (1984): 135–79; F. S. vom Saal, "Sexual Differentiation in Litter-Bearing Mammals: Influence of Sex of Adjacent Fetuses in Utero," *Journal of Animal Science* 67 (1989): 1824–40; Lydia W. Keisler et al., "Hormonal Manipulation of the Prenatal Environment Alters Reproductive Morphology and Increases Longevity in Autoimmune Nzb/W Mice," *Biology of Reproduction* 44 (1991): 707–16.

2. Susan C. Nagel et al., "Relative Binding Affinity-Serum Modified Access (Rba-Sma) Assay Predicts the Relative in Vivo Bioactivity of the Xenoestrogens Bisphenol A and Octyphenol," *Environmental Health Perspectives* 105 (1997): 70–76; Daniel M. Sheehan and Frederick S. vom Saal, "Low Dose Effects of Endocrine-Disruptors: A Challenge for Risk Assessment," *EPA's Risk Policy Report*, September 19, 1997, 31–39.

3. Vom Saal, "Intrauterine Position Phenomenon"; vom Saal and Bronson, "Sexual Characteristics."

4. Theo Colborn, Dianne Dumanoski, and John Peterson Myers, *Our Stolen Future: Are We Threatening Our Fertility, Intelligence, and Survival? A Scientific Detective Story* (New York: Penguin Books, 1996), 29–41.

5. Vom Saal, "Sexual Differentiation."

6. Ibid., 1836.

7. W. V. Welshons et al., "Large Effects from Small Exposures. I. Mechanisms for Endocrine-Disrupting Chemicals with Estrogenic Activity," *Environmental Health Perspectives* 111 (2003): 994–1006. Free hormone levels are used in clinical diagnosis.

8. University of Missouri, curriculum vitae of Wade V. Welshons, June 2001, http://endocrinedisruptors.missouri.edu/welshons/welshons.html (accessed March 24, 2010).

9. Benita S. Katzenellenbogen, "In Memoriam: Jack Gorski," *Endocrinology* 147 (2006): 5017–18.

10. National Research Council (NRC), Committee on Hormonally Active Agents in the Environment, "Dosimetry," in *Hormonally Active Agents in the Environment* (Washington, DC: National Academy Press, 1999), 82–118.

11. Katzenellenbogen, "In Memoriam"; J.R. Malayer and J. Gorski, "An Integrated Model of Estrogen Receptor Activity," *Domestic Animal Endocrinology* 10 (1993): 159–77.

12. Susan P. Porterfield, *Endocrine Physiology* (St. Louis, MO: Mosby, 1997), 7.

13. F.S. vom Saal and C. Hughes, "An Extensive New Literature Concerning Low Dose Effects of Bisphenol A Shows the Need for a New Risk Assessment," *Environmental Health Perspectives* 113 (2005): 926–33; Porterfield, *Endocrine Physiology,* 7.

14. NRC, "Dosimetry," 109–15.

15. Welshons et al., "Large Effects."

16. Joseph Thornton, "Nonmammalian Nuclear Receptors: Evolution and Endocrine Disruption," *Pure and Applied Chemistry* 75 (2003): 1827–39.

17. F.S. vom Saal et al., "Prostate Enlargement in Mice Due to Fetal Exposure to Low Doses of Estradiol or Diethylstilbestrol and Opposite Effects at High Doses," *Proceedings of the National Academy of Sciences* 94 (1997): 2056–61.

18. Ibid.; F.S. vom Saal et al., "A Physiologically Based Approach to the Study of Bisphenol A and Other Estrogenic Chemicals on the Size of Reproductive Organ, Daily Sperm Production and Behavior," *Toxicological and Industrial Health* 14 (1998): 239–60.

19. Fred vom Saal, interview by Doug Hamilton, producer of *Frontline,* "Fooling with Nature," February 1998, www.pbs.org/wgbh/pages/frontline/shows/nature/interviews/vomsaal.html.

20. Jerry J. Heindel, "Role of Exposure to Environmental Chemicals in the Developmental Basis of Reproductive Disease and Dysfunction," *Seminars in Reproductive Medicine* 24 (2006): 168–77; J.J. Heindel, "The Fetal Basis of Adult Disease: Role of Environmental Exposures—Introduction," *Birth Defects Research, Part A: Clinical and Molecular Teratology* 73 (2005): 131–32.

21. Z. Stein and M. Susser, "Effects of Early Nutrition on Neurological and Mental Competence in Human Beings," *Psychological Medicine* 15 (1985): 717–26; G.P. Ravelli, Z. Stein, and M.W. Susser, "Obesity in Young Man after Famine Exposure in Utero and Early Infancy," *New England Journal of Medicine* 295 (1976): 349–53; Z. Stein and M. Susser, "The Dutch Famine, 1944–1945, and the Reproductive Process. I. Effects or Six Indices at Birth," *Pediatric Research* 9 (1975): 70–76; Z. Stein and M. Susser, "The Dutch Famine, 1944–1945, and the Reproductive Process. II. Interrelations of Caloric Rations and Six Indices at Birth," *Pediatric Research* 9 (1975): 76–83. For a further discussion of the Barker thesis, see Heindel, "Fetal Basis."

22. Nagel et al., "Relative Binding Affinity"; vom Saal et al., "Physiologically Based Approach."

23. S. C. Nagel, F. S. vom Saal, and W. V. Welshons, "The Effective Free Fraction of Estradiol and Xenoestrogens in Human Serum Measured by Whole Cell Uptake Assays: Physiology of Delivery Modifies Estrogenic Activity," *Proceedings of the Society for Experimental Biology and Medicine* 217 (1998): 300–309; S. C. Nagel, F. S. vom Saal, and W. V. Welshons, "Developmental Effects of Estrogenic Chemicals Are Predicted by an in Vitro Assay Incorporating Modification of Cell Uptake by Serum," *Journal of Steroid Biochemistry and Molecular Biology* 69 (1999): 343–57.

24. José Antonio Brotons et al., "Xenoestrogens Released from Lacquer Coatings in Food Cans," *Environmental Health Perspectives* 103 (1995): 608–12; N. Olea et al., "Estrogenicity of Resin-Based Composites and Sealants Used in Dentistry," *Environmental Health Perspectives* 104 (1996): 289–305.

25. Vom Saal et al., "Prostate Enlargement in Mice."

26. E-mail from George Daston to EDSTAC [Endocrine Disruptor Screening and Testing Advisory Committee] listserv, subject: Re: Theo's Contribution to Listserv Dialogue, December 23, 1997, and e-mail from George Daston to EDSTAC listserv, subject: RE: Part II of report, June 11, 1998, both in Theo Colborn, personal papers, Paonia, CO, uncatalogued collection (hereafter cited as TC papers).

27. Christopher Borgert, phone interview by author, December 3, 2010.

28. Vom Saal, interview by Doug Hamilton.

29. Sean Milmo, "Chemical Sector Needs to Spend More to Study Endocrine Issues," *Chemical Market Reporter*, October 20, 1997, 7.

30. E-mail from Fred vom Saal to EDSTAC listserv, subject: Low dose workshop report, June 15, 1998, TC papers.

31. John Ashby, H. Tinwell, and J. Haseman, "Lack of Effects for Low Dose Levels of BPA and Diethylstilbestrol on the Prostate Gland of CF1 Mice Exposed in Utero," *Regulatory Toxicology and Pharmacology* 30 (1999): 156–66; S. Z. Cagen et al., "Normal Reproductive Organ Development in Wistar Rats Exposed to BP A in the Drinking Water," *Regulatory Toxicology and Pharmacology* 30 (1999): 130–39.

32. "Study Finds No Danger from Bisphenol-A," *Chemical Market Reporter* 254 (1998): 4; "New Study of BPA Finds No Negative Health Effects," *Chemical Market Reporter* 255 (1999): 23.

33. S. Z. Cagen et al., "Normal Reproductive Organ Development in CF-1 Mice Following Prenatal Exposure to Bisphenol A," *Toxicological Sciences* 50 (1999): 36–44.

34. James C. Lamb, "Can Today's Risk Assessment Paradigms Deal with Endocrine Active Chemicals?" *EPA's Risk Policy Report*, September 19, 1997, 30, 32–33, 39. Request for author interview with Lamb went unanswered.

35. Ibid.

36. Sheehan and vom Saal, "Low Dose Effects."

37. Pauli quoted in "FDA Unimpressed by Low Dose Claims," *Endocrine/Estrogen Letter*, May 20, 1999.

38. *Federal Register* 65, no. 4 (January 6, 2000): 784–87.

39. Ronald Melnick, National Institute of Environmental Health Sciences [NIEHS], Research Triangle Park, NC, personal communication to author, July 14, 2006.

40. Ibid.

41. NIH, NIEHS, National Toxicology Program [NTP], *National Toxicology Program's Report of the Endocrine Disruptors Low Dose Peer Review*, sponsored by the EPA and NIEHS/NTP, August 2001, http://ntp.niehs.nih .gov/?objectid=06F5CE98-E82F-8182–7FA81C02D3690D47. The studies coauthored by Waechter are Cagen et al., "Normal Reproductive Organ Development in Wistar Rats" and "Normal Reproductive Organ Development in CF-1 Mice."

42. *Federal Register* 65, no. 4 (January 6, 2000): 784–87.

43. R. Melnick et al., "Summary of the National Toxicology Program's Report of the Endocrine Disruptors Low Dose Peer Review," *Environmental Health Perspectives* 110 (2002): 427–31.

44. NTP, *National Toxicology Program's Report*, iv.

45. Ibid., vii.

46. Ibid., iv. In reviewing Rochelle Tyl's research, the panel confirmed the statistical analysis but found that the study minimized the significance of reported increases in uterine weights in animals exposed to bisphenol A (A-79).

47. Melnick et al., "Summary," 428–29.

48. All of the public comments submitted to the NTP are included in NTP, *National Toxicology Program's Report*, Appendix C.

49. G. P. Daston, J. C. Cook, and R. J. Kavlock, "Uncertainties for Endocrine Disrupters: Our View on Progress," *Toxicological Sciences* 74 (2003): 245–52.

50. NTP, *National Toxicology Program's Report*, C-23.

51. Ibid., C-47.

52. EPA, "EPA Statement Regarding Endocrine-Disruptor Low Dose Hypothesis," press release, March 26, 2002.

53. The Union of Concerned Scientists investigations found extensive scientific manipulation and misconduct within the Bush administration. Union of Concerned Scientists, "Scientific Integrity in Policymaking: An Investigation into the Bush Administration's Misuse of Science," July 2004, www .ucsusa.org/scientific_integrity/abuses_of_science/reports-scientific-integrity .html (accessed December 2, 2011).

54. NTP, *National Toxicology Program's Report*, C-52.

55. Richard C. Rue, vice president of the Annapolis Center, to Thomas J. Borelli, PhD, director of science and environmental policy, Phillip Morris Management Corp., February 11, 1999, document no. 2065243786; invoice no. 145 to Thomas J. Borelli, PhD, director of science and environmental policy, Phillip Morris Management Corp., "Corporate Strategic Sponsor," for $25,000, January 6, 1999, document no. 2065243787, Phillip Morris USA Documents, www.pmdocs.com/ (hereafter referred to as PM Documents); "Factsheet: Annapolis Center for Science-Based Public Policy, ACSBPP," n.d., Exxonsecrets.org, www.exxonsecrets.org/html/orgfactsheet.php?id=13

(accessed March 14, 2008). The environmental advocacy group Greenpeace organizes the website Exxonsecrets.org, with links to ExxonMobil's annual reports that list funders of the Annapolis Center. According to the website, the center has received just under $800,000 from the oil giant since 1998.

56. A history of the center is taken from its website from December 4, 2000, located through the "Wayback Machine," http://web.archive.org/web/20010202151700/www.annapoliscenter.org/history.htm (accessed November 3, 2010).

57. Richard C. Rue, vice president of the Annapolis Center, to Thomas J. Borelli, PhD, director of science and environmental policy, Phillip Morris Management Corp., January 29, 1998, document no. 2073684230, PM Documents. Reports released by the center include "Annapolis Accords for Cost-Benefit Analysis," "Epidemiology in Decision-Making," "How Can I Judge Good Science?" "Annapolis Accord for Risk Assessment," and the "Annapolis Accords on the Use of Toxicology in Risk Assessment and Decision-Making."

58. "Right to Know Regulatory X-Pert System," presentation by the Annapolis Center, October 1, 1998, document no. 2065243873, PM Documents.

59. Gray did not recall the system. George Gray, conversation with author, November 11, 2010. Hahn did not respond to an e-mail request for interview with author.

60. Annapolis Center for Science-Based Public Policy, "Strategic Planning Committee Draft Discussion Piece," October 29, 1998, 13, document no. 2065243880, PM Documents.

61. F.W. Lipfert, "The 'Particle Wars' and a Path to Peace," report by Annapolis Center for Science-Based Public Policy, 2003. The study is discussed in David Michaels, *Doubt Is Their Product: How Industry's Assault on Science Threatens Your Health* (Oxford: Oxford University Press, 2008), 56–57.

62. Annapolis Center for Science-Based Public Policy, "Strategic Planning Committee."

63. Participants included representatives of other risk-analysis institutes, including the Gradient Corporation and HealthRisk Strategies, directed by Gail Charnley, formerly with the Harvard Center for Risk Analysis.

64. In a conversation with the author, Gray could not recall if he was serving on the Annapolis Center board when he received funding to coordinate the forum. According to the Annapolis Center website at the time, he was listed as a board member. George Gray, conversation with author, November 11, 2010.

65. George M. Gray et al., "The Annapolis Accords on the Use of Toxicology in Risk Assessment and Decision-Making: An Annapolis Center Workshop Report," *Toxicology Methods* 11 (2001): 225–31.

66. George M. Gray et al., "Weight of the Evidence Evaluation of Low Dose Reproductive and Developmental Effects of Bisphenol A," *Human and Ecological Risk Assessment* 10 (2004): 876.

67. NTP, *National Toxicology Program's Report*, i.

68. Gray et al., "Weight of the Evidence Evaluation," 876–77.

69. Gerald E. Markowitz and David Rosner, *Deceit and Denial: The Deadly Politics of Industrial Pollution* (Berkeley: University of California Press, 2002), 215. The GLP standard was developed as legal guidelines for research conducted by private laboratories under government and industry contract in response to evidence of falsified data, foul laboratory conditions, and overall sloppy work and research at the laboratory Industrial Bio-Test in the early 1980s.

70. Gray et al., "Weight of the Evidence Evaluation," 876–77.

71. M. Ema et al., "Rat Two-Generation Reproductive Toxicity Study of Bisphenol A," *Reproductive Toxicology* 15, no. 5 (2001): 505–23; R. W. Tyl et al., "Three-Generation Reproductive Toxicity Study of Dietary Bisphenol A in Cd Sprague-Dawley Rats," *Toxicological Sciences* 68 (2002): 121–46.

72. Gray et al., "Weight of the Evidence Evaluation," 909.

73. Ibid., 875.

74. Ibid.

75. Ibid., 878.

76. A. B. Hill, "The Environment and Disease: Association or Causation?" *Proceedings of the Royal Society of Medicine* 58 (1965): 298.

77. Gray et al., "Weight of the Evidence Evaluation," 911.

78. R. J. Witorsch, "Low-Dose in Utero Effects of Xenoestrogens in Mice and Their Relevance to Humans: An Analytical Review of the Literature," *Food and Chemical Toxicology* 40 (2002): 905–12.

79. Kenneth L. Becker et al., eds., *Principles and Practice of Endocrinology and Metabolism*, 2nd ed. (Philadelphia: J. B. Lippincott Company, 1995), 988.

80. R. J. Witorsch, "Endocrine Disruptors: Can Biological Effects and Environmental Risks Be Predicted?" *Regulatory Toxicology and Pharmacology* 36 (2002): 118–30; R. J. Witorsch, "Low Dose in Utero Effects."

81. Witorsch's name appeared on a list of experts available through Juris-Pro Expert Witness Testimony, www.jurispro.com/RaphaelWitorsch (accessed September 14, 2007).

82. "Comments on the External Review Draft (950300) Entitled 'Developmental and Reproductive Effects of Exposure to Environmental Tobacco Smoke.' Submitted to: California Environmental Protection Agency Office of Environmental Health Hazard Assessment (OEHHA) Science Advisory Board's Developmental and Reproductive Toxicant (DART) Identification Committee, Raphael J. Witorsch, Ph.D., Professor of Physiology, School of Medicine, Medical College of Virginia of Virginia Commonwealth University, Richmond, Virginia, and Philip Witorsch, M.D., Professor of Medicine and Pharmacology, Georgetown University Medical Center, Washington, D.C. These comments are being submitted at the request and with the support of the law firm of Covington and Burling, Washington, D.C., on behalf of their client, the Tobacco Institute, Washington, D.C., Tobaccodocuments.org, http://tobaccodocuments.org/bw/1482222.html (accessed September 26, 2007).

83. J. E. Heinze, "Adverse Health Effects of Bisphenol A in Early Life," *Environmental Health Perspectives* 111 (2003): A382–83; author reply, A83; Liz Szabo, "Non-Stick Chemicals May Cut Birth Weight; Two New Studies

Add to Cookware Scrutiny," *USA Today,* August 23, 2007, 7D; Center for Science in the Public Interest, Integrity in Science project, www.cspinet .org/integrity/nonprofits/environmental_health_research_foundation.html (accessed September 18, 2007).

84. John Adams Associates, Inc., "The JAA Team," 2010, www.johnadams .com/ourteam.php.

85. John D. Graham and Jonathan Baert Wiener, eds., *Risk versus Risk: Tradeoffs in Protecting Health and the Environment* (Cambridge, MA: Harvard University Press, 1995); John D. Graham and Jonathan Baert Wiener, "Confronting Risk Tradeoffs," in Graham and Wiener, *Risk versus Risk,* 1–41.

86. Vincent Ostrom and Elinor Ostrom, "Public Goods and Public Choices," in *Alternatives for Delivering Public Services: Toward Improved Performance,* ed. E. S. Savas (Boulder, CO: Westview Press, 1977), 7–49; Susan Rose-Ackerman, *Rethinking the Progressive Agenda: The Reform of the American Regulatory State* (New York: Free Press, 1992).

87. Mary Douglas and Aaron B. Wildavsky, *Risk and Culture: An Essay on the Selection of Technical and Environmental Dangers* (Berkeley: University of California Press, 1982).

88. A. Stirling and D. Gee, "Science, Precaution, and Practice," *Public Health Reports* 117 (2002): 521–33.

89. John D. Graham, communication to author, November 4, 2010.

90. "Faculty Directory: John D. Graham," biography, Frederick S. Pardee RAND Graduate School, www.indiana.edu/~spea/faculty/graham-johnd .shtml (accessed December 30, 2011).

91. Chris Mooney, "Some Like It Hot," *Mother Jones,* May–June 2005, 36–94.

92. Anne Barnard, "Group Blasts Bush Nominee for Industry-Tied Research," *Boston Globe,* March 13, 2001, A2.

93. Lists of members on the executive council were obtained through the Way Back Machine, www.web.archive.org (accessed June 19, 2008); King and Spalding, LLP, *http://www.kslaw.com/.*

94. These are some of the leading scholars of risk science listed by Graham and Wiener in "Confronting Risk Tradeoffs," 18. See Cass Sunstein, *Risk and Reason: Safety, Law and the Environment* (Cambridge: Cambridge University Press, 2002). On Huber, see Sheila Jasanoff, *Science at the Bar: Law, Science, and Technology in America* (Cambridge, MA: Harvard University Press, 1995), 12–13. Graham notes that Huber favored federal regulation over "a plethora of state rules or a morass of tort liability suits before juries" (Graham, communication to author, November 4, 2010). On Wildavsky, see Douglas and Wildavsky, *Risk and Culture;* Mary Douglas, *Risk and Blame: Essays in Cultural Theory* (London: Routledge, 1992). On Ames, see Robert Proctor, "Natural Carcinogens and the Myth of Toxic Hazards," in *Cancer Wars: How Politics Shapes What We Know and Don't Know about Cancer* (New York: Basic Books, 1995). In the 1970s, Ames supported environmental concerns about carcinogenic pesticides. After being encouraged to read the writing of

Friedrich von Hayek, Ames experienced a political transformation to market liberalism. In the late 1980s, Ames developed the Human Exposure/Rodent Potency Index, which ranked carcinogens, natural and synthetic, according to their carcinogenic potency. The objective behind the index was to demonstrate that not all natural chemicals were benign nor, conversely, were all synthetic chemicals toxic. Ames became an outspoken critic of environmentalism and the regulation of chemicals.

95. Sunstein, *Risk and Reason*, 12–18; George Gray and John D. Graham, "Regulating Pesticides," in Graham and Wiener, *Risk versus Risk*, 173–92.

96. Ulrich Beck, *Risk Society: Towards a New Modernity* (Newbury Park, CA: Sage Publications, 1992).

97. David Pellow, "Environmental Inequality Formation: Toward a Theory of Environmental Injustice," *American Behavioral Scientist* 43 (2000): 581–601; Robert D. Bullard, *Dumping in Dixie: Race, Class, and Environmental Quality* (Boulder, CO: Westview Press, 1990); Robert D. Bullard, *Unequal Protection: Environmental Justice and Communities of Color* (San Francisco: Sierra Club Books, 1994).

98. Frank Ackerman, *Poisoned for Pennies: The Economics of Toxins and Precaution* (Washington, DC: Island Press, 2008); Christopher H. Schroeder, Rena Steinzor, and Center for Progressive Regulation (U.S.), *A New Progressive Agenda for Public Health and the Environment: A Project of the Center for Progressive Regulation* (Durham, NC: Carolina Academic Press, 2005).

99. C. F. Cranor, "Some Legal Implications of the Precautionary Principle: Improving Information-Generation and Legal Protections," *International Journal of Occupational and Environmental Health* 17 (2004): 17–34; P. L. deFur and M. Kaszuba, "Implementing the Precautionary Principle," *Science of the Total Environment* 288 (2002): 155–65; Carolyn Raffensperger and Joel A. Tickner, eds., *Protecting Public Health and the Environment: Implementing the Precautionary Principle* (Washington, DC: Island Press, 1999); Ken Geiser, "Establishing a General Duty of Precaution in Environmental Protection Policies in the United States," in Raffensperger and Tickner, *Protecting Public Health*, xxi–xxvi; Joel Tickner, Carolyn Raffensperger, and Nancy Myers, "The Precautionary Principle in Action: A Handbook," written for the Science and Environmental Health Network, 1998, www.biotech-info.net/handbook.pdf; Sheldon Krimsky, "The Precautionary Approach," *Forum for Applied Research and Public Policy* 13 (1998): 34–37.

100. Stirling and Gee, "Science, Precaution, and Practice."

101. John D. Graham, "The Perils of the Precautionary Principle: Lessons from the American and European Experience," lecture delivered as part of Heritage Lectures at the Heritage Foundation, October 20, 2003, Heritage Foundation, www.heritage.org/Research/Regulation/hl818.cfm; Chris Mooney, *The Republican War on Science* (New York: Basic Books, 2005), 115; Public Citizen, "Industry Hired Gun, Hostile to Health and Environmental Safeguards, Up for Key Regulatory Post in Bush Administration," press release, March 12, 2001.

102. Gray was confirmed as assistant administrator in 2005 and served until 2009; *http://epa.gov/ncer/events/calendar/2006/apr26/bios.html.*

103. J. E. Goodman et al., "An Updated Weight of the Evidence Evaluation of Reproductive and Developmental Effects of Low Doses of Bisphenol A," *Critical Reviews in Toxicology* 36 (2006): 387–457. George Gray informed the author that the intent of the criteria used in the weight of the evidence methodology was not to inform the "reliability" or "relevancy" of the data as it related to the rules of evidence, saying that the selection of the terms was happenstance. Gray, conversation with author, November 11, 2010.

104. Sheila Jasanoff, "Law's Knowledge: Science for Justice in Legal Settings," *American Journal of Public Health* 95, suppl. 1 (2005): S52.

105. Thomas O. McGarity and Wendy Wagner, *Bending Science: How Special Interests Corrupt Public Health Research* (Cambridge, MA: Harvard University Press, 2008), 26.

106. Tozzi quoted in Mooney, *Republican War on Science*, 105; Rick Weiss, " 'Data Quality' Law Is Nemesis of Regulation," *Washington Post*, August 16, 2004, A1.

107. Mooney, *Republican War on Science*, 103.

108. "Jim Tozzi Views on Centralized Regulatory Review Contained in the National Archives Interview," 2009, Center for Regulatory Effectiveness, www.thecre.com/video/National_Archive.html.

109. Paperwork Reduction Act, 44 USC 3501 (1980).

110. Shelley Lynne Tomkin, *Inside OMB: Politics and Process in the President's Budget Office* (Armonk, NY: M. E. Sharpe, 1998), 207–08.

111. Weiss, " 'Data Quality' Law."

112. Tomkin, *Inside OMB*, 208.

113. Tozzi quoted in Mooney, *Republican War on Science*, 105.

114. *Dole v. United Steelworkers of America*, 494 US 26 (1990).

115. Jim Tozzi, Multinational Business Services, to Dr. Elizabeth Fontham, Louisiana State Medical Center, April 14, 1997, document no. 2065345898, PM Documents.

116. Omnibus Appropriations Act, Public Law 105–277 (1999).

117. Mooney, *Republican War on Science*, 103; Weiss, " 'Data Quality' Law"; Michaels, *Doubt Is Their Product*, 176–77; Roni A. Neff and Lynn R. Goldman, "Regulatory Parallels to *Daubert*: Stakeholder Influence, 'Sound Science,' and the Delayed Adoption of Health-Protective Standards," *American Journal of Public Health* 95, suppl. 1 (2005): S85.

118. Jim Tozzi, Center for Regulatory Effectiveness, to Dr. Robert Elves, Phillip Morris USA, January 24, 2000, document no. 2072826845, and Jim Tozzi, Center for Regulatory Effectiveness, to Dr. Robert Elves, Phillip Morris, February 29, 2000, document no. 2072826830, both in PM Documents.

119. Treasury and General Government Appropriation Act for Fiscal Year 2001, Public Law 106–554, Sec. 515, 106th Cong. (2000), www.fws.gov/informationquality/section515.html (accessed May 27, 2008).

120. McGarity and Wagner, *Bending Science*, 151.

121. Ibid.; Mooney, *Republican War on Science,* 108–15; Weiss, " 'Data Quality' Law." Tozzi drafted a petition using the Data Quality Act to challenge the findings of Tyrone Hayes's atrazine research, which reported amphibian deformities from exposure to the herbicide, in 2002, when the EPA was deciding whether to restrict use of the herbicide. Tozzi's petition helped to delay the EPA's action on atrazine.

122. Tozzi quoted in Mooney, *Republican War on Science,* 151.

123. Tozzi quoted in Weiss, " 'Data Quality' Law."

124. Mooney, *Republican War on Science,* 116–19.

125. Jocelyn Kaiser, "New Regulatory Czar Takes Charge," *Science* 294 (2001): 32–33. It is important to note that Graham's position is not simply antiregulatory. For example, in 2001 his office pushed the FDA to require labels on foods with trans-fatty acids.

126. Mooney, *Republican War on Science,* 117–19.

127. NRC, National Academy of Sciences, Committee to Review the OMB Risk Assessment Bulletin, *Scientific Review of the Proposed Risk Assessment Bulletin from the Office of Management and Budget* (Washington, DC: National Academy Press, 2007), 6.

128. Terry Quill, remarks at Villanova Environmental Law Journal Symposium, "Low-Dose Toxicity: Scientific Controversies, Regulatory Efforts and Potential Litigation," Villanova, PA, October 6, 2007; author observed. At the conference, Quill outlined a number of criteria for conducting a weight-of-the-evidence assessment, repeating verbatim several of the Annapolis Accords standards without, however, directly naming the accords. By contrast, he argued that consensus statements did not reflect evaluations of sound science.

129. Michaels, *Doubt Is Their Product,* 54.

130. Quill, remarks at Villanova Environmental Law Journal Symposium. The question was posed by the author during a question-and-answer session.

131. "DuPont Won't Face Criminal Charges over Allegations of Hiding Information," *FinancialWire,* October 16, 2007, 1.

132. Reps. John Dingell and Bart Stupak, Comm. on Energy and Commerce, to Matthew R. Weinberg, CEO, The Weinberg Group, February 5, 2008. The 2003 letter was included in the congressmen's letter of 2008. P. Terrence Gaffney, vice president, Product Defense, The Weinberg Group Inc., to Jane Brooks, vice president, Special Initiative, DuPont de Nemours & Company, April 29, 2003.

133. "DuPont Won't Face Criminal Charges," 1.

134. Privately, several researchers mentioned to me that they had stepped away from the issue because it became so controversial and personal.

135. Vom Saal and Hughes, "Extensive New Literature."

136. Ibid.

137. Ibid.

138. C.M. Markey et al., "In Utero Exposure to Bisphenol A Alters the Development and Tissue Organization of the Mouse Mammary Gland," *Biology of Reproduction* 65 (2001): 58.

139. S.M. Ho et al., "Developmental Exposure to Estradiol and Bisphenol A Increases Susceptibility to Prostate Carcinogenesis and Epigenetically Regulates Phosphodiesterase Type 4 Variant 4," *Cancer Research* 66 (2006): 5624–32; G. S. Prins et al., "Developmental Estrogen Exposures Predispose to Prostate Carcinogenesis with Aging," *Reproductive Toxicology* 23 (2007): 374–82.

140. P.A. Hunt et al., "Bisphenol A Exposure Causes Meiotic Aneuploidy in the Female Mouse," *Current Biology* 13 (2003): 546–53; P.A. Hunt et al., "The Bisphenol A Experience: A Primer for the Analysis of Environmental Effects on Mammalian Reproduction," *Biology of Reproduction* 81 (2009): 807–13.

141. W. Volkel et al., "Metabolism and Kinetics of Bisphenol A in Humans at Low Doses Following Oral Administration," *Chemical Research in Toxicology* 15 (2002): 1281–87.

142. A.M. Calafat et al., "Urinary Concentrations of Bisphenol A and 4-Nonylphenol in a Human Reference Population," *Environmental Health Perspectives* 113 (2005): 391–95.

143. A. W. Schwartz and P. J. Landrigan, "Bisphenol A in Thermal Paper Receipts: An Opportunity for Evidence-Based Prevention," *Environmental Health Perspectives* 120 (2012): a14-a15.

144. Mark Kirschner, "Chemical Profile: Bisphenol-A," *Chemical Market Reporter*, December 20–27, 2004.

6. BATTLES OVER BISPHENOL A

1. Lindsay Layton, "No BPA for Baby Bottles in U.S.," *Washington Post*, March 6, 2009, A6.

2. Matthew Perrone, "Sunoco Restricts Sales of Chemicals Used in Bottles," *Washington Post*, March 12, 2009.

3. Valerie Jablow, "Chemical in Plastic Bottles Fuels Science, Concern—and Litigation," *Trial* 44, no. 8 (2008): 12.

4. Long-Range Research Initiative, American Chemistry Council, LRI Research Strategy 2009–2015: Modernizing Approaches to Chemical Risk Assessment, 2009, www.americanchemistry.com/s_acc/sec_lri .asp?CID=1389&DID=5073.

5. National Research Council (NRC), Committee on Hormonally Active Agents in the Environment, "Dosimetry," in *Hormonally Active Agents in the Environment* (Washington, DC: National Academy Press, 1999), 82–118.

6. "Program Announcement (PA) Title: The Fetal Basis of Adult Disease: Role of the Environment," PAR-02–105, release date May 2, 2002, National Institute of Environmental Health Sciences [NIEHS], http://grants.nih.gov/ guide/pa-files/PAR-02–105.html.

7. Jerry Heindel, Division of Extramural Research and Training (DERT), Cellular, Organ, and Systems Pathobiology, interview by author, Research Triangle Park, NC, July 18, 2006.

8. W. A. Toscano and K. P. Oehlke, "Systems Biology: New Approaches to Old Environmental Health Problems," *International Journal of Environmental Research and Public Health* 2, no. 1 (2005): 4–9.

9. D. Crews and J. A. McLachlan, "Epigenetics, Evolution, Endocrine Disruption, Health, and Disease," *Endocrinology* 147, no. 6, suppl. (2006): S4–10; Ethan Watters, "DNA Is Not Destiny: The New Science of Epigenetics Rewrites the Rules of Disease, Heredity, and Identity," *Discover Magazine*, November 22, 2006, 32.

10. Crews and McLachlan, "Epigenetics, Evolution," S4.

11. Ibid., S6.

12. M. D. Anway et al., "Epigenetic Transgenerational Actions of Endocrine Disruptors and Male Fertility," *Science* 308 (2005): 1466–69.

13. Jerry Heindel, Division of Extramural Research and Training (DERT), Cellular, Organ, and Systems Pathobiology, interview by author, Research Triangle Park, NC, July 18, 2006.

14. "Program Announcement (PA) Title: The Fetal Basis of Adult Disease: Role of the Environment."

15. G. S. Prins et al., "Developmental Estrogen Exposures Predispose to Prostate Carcinogenesis with Aging," *Reproductive Toxicology* 23 (2007): 374–82.

16. Gail Prins, personal communication to author, June 2, 2010.

17. S. M. Ho et al., "Developmental Exposure to Estradiol and Bisphenol A Increases Susceptibility to Prostate Carcinogenesis and Epigenetically Regulates Phosphodiesterase Type 4 Variant 4," *Cancer Research* 66 (2006): 5624–32.

18. "Bisphenol A: An Expert Panel Examination of the Relevance of Ecological, in Vitro and Laboratory Animal Studies for Assessing Risks to Human Health," NIEHS, National Institute of Dental and Craniofacial Research [NIDCR], EPA, Commonweal, November 28–29, 2006; author attended entire meeting.

19. "Bisphenol A: An Expert Panel Examination of the Relevance of Ecological, in Vitro, and Laboratory Animals Studies for Assessing Risk to Human Health," Chapel Hill, NC, November 28–29, 2006; author observed both days. An official report on the meeting's organization and conclusions was published the following year. F. S. vom Saal et al., "Chapel Hill Bisphenol A Expert Panel Consensus Statement: Integration of Mechanisms, Effects in Animals and Potential to Impact Human Health at Current Levels of Exposure," *Reproductive Toxicology* 24 (2007): 131–38.

20. Vom Saal et al., "Chapel Hill Bisphenol A Expert Panel," 133.

21. Ibid. It should be noted that not all of the meeting's participants signed on to the final report. No minority report was drafted or discussed.

22. Brendan Borrell, "Toxicology: The Big Test for Bisphenol A," *Nature* 464 (2010): 1122–24; J. P. Myers et al., "Why Public Health Agencies Cannot Depend on Good Laboratory Practices as a Criterion for Selecting Data: The Case of Bisphenol A," *Environmental Health Perspectives* 117 (2009): 309–15.

23. Devra Lee Davis, *When Smoke Ran Like Water: Tales of Environmental Deception and the Battle against Pollution* (New York: Basic Books, 2002); Dan Fagin, Marianne Lavelle, and Center for Public Integrity, *Toxic Deception: How the Chemical Industry Manipulates Science, Bends the Law, and Endangers Your Health* (Secaucus, NJ: Carol Publishing Group, 1996); Robert Gottlieb, *Forcing the Spring: The Transformation of the American Environmental Movement* (Washington, DC: Island Press, 1993); Samuel P. Hays and Barbara D. Hays, *Beauty, Health, and Permanence: Environmental Politics in the United States, 1955–1985*, Studies in Environment and History (Cambridge [Cambridgeshire] and New York: Cambridge University Press, 1987); Gerald E. Markowitz and David Rosner, *Deceit and Denial: The Deadly Politics of Industrial Pollution* (Berkeley: University of California Press, 2002); David Rosner and Gerald E. Markowitz, *Deadly Dust: Silicosis and the Politics of Occupational Disease in Twentieth-Century America* (Princeton: Princeton University Press, 1991).

24. Indeed, speaking about toxic chemicals seems to demand thick skin. At one event, I received a public berating by a high-level chemical industry official when presenting some of this history. At an international scientific meeting on endocrine disruption, many researchers privately shared with me their own experiences of being harassed by industry officials.

25. David Case, "Warning: This Bottle May Contain Toxic Chemicals. Or Not," *Fast Company*, February 2009, 90–99.

26. Marla Cone, "Public Health Agency Linked to Chemical Industry," *Los Angeles Times*, March 4, 2007, 16.

27. Case, "Warning."

28. Bette Hileman, "Probe of NIEHS Head Broadens," *Chemical and Engineering News*, September 3, 2007, 9; Rick Weiss, "Probe Finds NIH Official Violated Government Regulations," *Washington Post*, June 27, 2007, A3.

29. Cone, "Public Health Agency"; Richard Wiles, executive director, Environmental Working Group, to David Schwartz, director, National Toxicology Program [NTP], NIEHS, February 28, 2007, www.ewg.org/node/21982; Henry A. Waxman, chairman, House Oversight and Government Reform Comm., and Barbara Boxer, chairman, Senate Environment and Public Works Comm., to David Schwartz, director, NIEHS, February 28, 2007, www.ewg.org/files/ltr_WaxmanBoxer.pdf.

30. Anila Jacob, senior scientist, Environmental Working Group, to Michael Shelby, director, Center for the Evaluation of Risks to Human Reproduction [CERHR], June 20, 2007, http://ntp.niehs.nih.gov/ntp/ohat/bisphenol/pubcomm/EWG_Comments_BPA_Interim.pdf.

31. Ronald Melnick, NIEHS, Research Triangle Park, NC, personal communication to author, July 14, 2006. Indeed, some risk analysts, notably Roger Cook, an expert on quantitative risk and uncertainty analysis at the Washington, D.C., think tank Resources for the Future (RFF), have developed complex

quantitative processes for selecting experts. R. M. Cooke and L. H. J. Goossens, "Expert Judgment Elicitation for Risk Assessment of Critical Infrastructures," *Journal of Risk Research* 7 (2004): 643–56.

32. Sharon H. Kneiss, V.P., Product Division, ACC [American Chemistry Council], to George M. Gray, assistant administrator, EPA, May 3, 2007; George Gray, assistant administrator, EPA, to Nancy Sandrof, manager, BFRIP [Brominated Flame Retardants Industry Panel], ACC, January 8, 2008; Environmental Working Group, "EPA Axes Panel Chair at Request of Chemical Industry Lobbyists: Review Panel Timeline," 2008, www.ewg.org/node/26033; H. Josef Hebert, "Lawmakers Probe EPA Conflicts," Associated Press, March 17, 2008.

33. NIEHS, NTP, NIH, Dept. of Health and Human Services, "Audit of Literature Cited and Fidelity of Requested Changes to Draft Bisphenol A Expert Panel Reports," July 24, 2007, http://ntp.niehs.nih.gov/ntp/ohat/bisphenol/SIauditreviewreportv12072407.pdf. Author attended the CERHR public meeting on the BPA review, CERHR Bisphenol A Expert Panel Meeting II, August 6–8, 2007, Hilton Hotel, Old Town Alexandria, VA.

34. A. W. Schwartz and P. J. Landrigan, "Bisphenol A in Thermal Paper Receipts: An Opportunity for Evidence-Based Prevention," *Environmental Health Perspectives* 120 (2012): a14-a15.

35. G. S. Prins et al., "Serum Bisphenol A Pharmacokinetics and Prostate Neoplastic Responses Following Oral and Subcutaneous Exposures in Neonatal Sprague-Dawley Rats," *Reproductive Toxicology* 31 (2010): 1–9, reported significant effects on prostate development and precancerous lesions in animals exposed by either injection or oral routes of exposure. They also reported that circulating levels of active BPA after injection (10 µg/kg) were similar to human biomonitoring levels. After several hours, levels of BPA in the blood were not statistically different by route of exposure.

36. CERHR, "NTP-CERHR Expert Panel Report on the Reproductive and Developmental Toxicity of Bisphenol A," November 26, 2007.

37. Laura N. Vandenberg et al., "Response to the Interim Draft of the NTP-CERHR Report on the Reproductive and Developmental Toxicity of Bisphenol A," public comments, June 20, 2007.

38. "Sound Science Prevails in Review of Bisphenol A," PR Newswire, August 8, 2007.

39. U.S. Department of Health and Human Services, NTP, CERHR, "NTP-CERHR Monograph on the Potential Human Reproductive and Developmental Effects of Bisphenol A," NIH Publication no. 08–5994, 2008.

40. Ibid., 38.

41. Ibid.

42. Rick Smith and Bruce Lourie, *Slow Death by Rubber Duck: The Secret Danger of Everyday Things* (Berkeley, CA: Counterpoint, 2009).

43. Ted Boadway, chairman of the federal science panel for the Canadian Chemical Management Program, interview by author, December 4, 2008.

44. "Government of Canada Protects Families with Bisphenol A Regulation," Health Canada, press release, October 17, 2008, www.chemicalsubstanc eschimiques.gc.ca/challenge-defi/batch-lot-2/bisphenol-a/index-eng.php.

45. Meg Kissinger and Susanne Rust, "BPA Industry Fights Back," *Milwaukee Journal Sentinel,* August 22, 2009, www.jsonline.com/watchdog/ watchdogreports/54195297.html.

46. "Class Action Lawsuit Filed against Baby Bottle Manufacturer," *Market Wire,* March 12, 2007. The suit filed against five major polycarbonate baby bottle producers was brought by Robert Weiss, a tort litigation lawyer who represented employees of WorldCom who lost their jobs and pensions in the company's accounting scandal in the early 2000s. The case argues that the class plaintiff was exposed to unsafe levels of bisphenol A that resulted in abnormal genital development.

47. Rep. John D. Dingell, chairman, Comm. on Energy and Commerce, "Committee to Investigate Chemical in Infant Formula Liners," press release, January 17, 2008.

48. Reps. John Dingell and Bart Stupak, Comm. on Energy and Commerce, to Ross Products Division, PBM Products, Nestle, Wyeth Nutrition, Solus Products, and Mead Johnson and Company, January 17, 2008, in author's files.

49. Layton, "No BPA."

50. Donald Kennedy, "Toxic Dilemmas," *Science* 318 (2007): 1217; Arlene Blum, "The Fire Retardant Dilemma," *Science* 318 (2007): 194.

51. Janet Nudelman, policy director, Breast Cancer Fund, personal communication to author, June 29, 2010.

52. Jon Entine, "The Troubling Case of Bisphenol A: At What Point Should Science Prevail?" *American Enterprise Institute for Public Policy Regulation Outlook,* no. 2, March 31, 2010, www.aei.org/article/health/ the-troubling-case-of-bisphenol-a/.

53. Environmental Health News, www.environmentalhealthnews.org/ (accessed September 21, 2011). The website was established in 2003 by John Peterson Myers, coauthor of *Our Stolen Future,* as a source of aggregated daily news reports on environmental health as well as summaries of new research.

54. Bill Moyers Journal, "Chemicals in Food," May 23, 2008, www.pbs .org/moyers/journal/05232008/watch2.html.

55. Meg Kissinger, "BPA Substitutes Sought," *Milwaukee Journal Sentinel,* November 1, 2008; Susanne Rust and Meg Kissinger, "BPA Leaches from 'Safe' Products," *Milwaukee Journal Sentinel,* November 15, 2008; Kissinger and Rust, "BPA Industry Fights Back"; Susanne Rust and Meg Kissinger, "FDA Relied Heavily on BPA Lobby," *Milwaukee Journal Sentinel,* May 16, 2009. The *Milwaukee Journal Sentinel* hosts a website, Chemical Fallout, with links to all of its articles on BPA and additional information: www.jsonline. com/watchdog/34405049.html.

56. Scott Openshaw, director, communications, Grocery Manufacturers Association [GMA], to GMA Communications Committee, GMA Federal Affairs Committee, GMA State Affairs Committee, "NAMPA," memoran-

dum, May 27, 2009, obtained by the Environmental Working Group and posted online, www.ewg.org/files/BPA-Joint-Trade-Association.pdf; Lindsay Layton, "Strategy Being Devised to Protect Use of BPA," *Washington Post,* May 31, 2009, A2.

57. Mia Davis et al., "Baby's Toxic Bottle: Bisphenol A Leaching from Popular Baby Bottles" (Work Group for Safe Markets, 2008); Carly Weeks, "Test Results Fuel Boom in Sales of Bisphenol A-Free Baby Bottles," *Globe and Mail,* February 14, 2008, www.theglobeandmail.com/life/article667837.ece.

58. Mike Belliveau, "Healthy States: Protecting Families from Toxic Chemicals while Congress Lags Behind," report sponsored by SAFER States, Safer Chemicals, Health Families Coalition, November 2010, http://blog.saferchemicals.org/2010/11/healthy-states-protecting-families-from-toxic-chemicals-while-congress-lags-behind.html.

59. Janet Nudelman, policy director, Breast Cancer Fund, personal communication to author, June 29, 2010.

60. In 2008, in response to a number of product recalls, Congress passed the Consumer Product Safety Improvement Act, which expanded the authority of the Consumer Product Safety Commission, banned lead from children's products, and phased out a number of phthalates from children's toys. Consumer Product Safety Commission, "The Consumer Product Safety Improvement Act (CPSIA) of 2008," n.d., www.cpsc.gov/about/cpsia/cpsia.html (accessed September 23, 2010).

61. The title of the California law provoked some debate by those concerned that the use of the phrase "green chemistry" was being conflated with chemical regulation. M. P. Wilson and M. R. Schwarzman, "Toward a New U.S. Chemicals Policy: Rebuilding the Foundation to Advance New Science, Green Chemistry, and Environmental Health," *Environmental Health Perspectives* 117 (2009): 1202–9; Karen Peabody O'Brien, John Peterson Myers, and John Warner, "Green Chemistry: Terminology and Principles," *Environmental Health Perspectives* 117 (2009): A385–86.

62. Green Chemistry Initiative, AB 1879, passed February 8, 2008, www.leginfo.ca.gov/pub/07–08/bill/asm/ab_1851–1900/ab_1879_bill_20080929_chaptered.html.

63. Sec. 2601–2692 [15 USC], Toxic Substances Control Act (TSCA) of 1976.

64. U.S. General Accounting Office [GAO], "Toxic Substances Control Act: Legislative Changes Could Make the Act More Effective," GAO/RCED-94–103, September 26, 1994; GAO, "Chemical Regulation Options Exist to Improve EPA's Ability to Assess Health Risks and Manage Its Chemical Review Program," GAO-05–458, June 2005.

65. *Corrosion Proof Fittings v. EPA,* 947 F.2d 1201 (5th Cir. 1991).

66. Sarah A. Vogel and Jody A. Roberts, "Why the Toxic Substances Control Act Needs an Overhaul, and How to Strengthen Oversight of Chemicals in the Interim," *Health Affairs* (May 30, 2011): 898–905.

67. Richard A. Denison, "Ten Essential Elements in TSCA Reform," *Environmental Law Reporter* 39 (2009): 10020–28.

68. David E. Adelman, "A Cautiously Pessimistic Appraisal of Trends in Toxics Regulation," *Journal of Law and Policy* 32 (2010): 1–44.

69. Denison, "Ten Essential Elements"; GAO, "Toxic Substances Control Act"; GAO, "Options for Enhancing the Effectiveness of the Toxic Substances Control Act," GAO-09-428T, 2009; John S. Applegate, "Bridging the Data Gap: Balancing the Supply and Demands for Chemical Information," *Texas Law Review* 86 (2008): 1365–1407.

70. GAO, "Report to the Congress: High-Risk Series: An Update," GAO-09-271, 2009.

71. Mark Schapiro, *Exposed: The Toxic Chemistry of Everyday Life and What's at Stake for American Power* (White River Junction, VT: Chelsea Green, 2007), 148.

72. ACC, "10 Principles for Modernizing TSCA," www.americanchemistry .com/Policy/Chemical-Safety/TSCA/10-Principles-for-Modernizing-TSCA .pdf (accessed January 12, 2011).

73. John Hontelez, ed., "Navigating REACH," report sponsored by European Environmental Bureau, Friends of the Earth—Europe, Greenpeace, 2007.

74. Schapiro, *Exposed*, 143–56.

75. Richard Denison, "Not That Innocent: A Comparative Analysis of Canadian, European Union and United States Policies on Industrial Chemicals," report sponsored by Environmental Defense Fund in cooperation with Pollution Probe, 2007, www.edf.org/sites/default/files/6149_NotThatInnocent_ Fullreport.pdf.

76. Isabelle Laborde, "REACH: The New European Union Chemicals Regulations," *Natural Resources and Environment* 23, no. 3 (2009): 63–65; C. Ruden and S. O. Hansson, "Registration, Evaluation, and Authorization of Chemicals (REACH) Is but the First Step—How Far Will It Take Us? Six Further Steps to Improve the European Chemicals Legislation," *Environmental Health Perspectives* 118 (2010): 6–10.

77. C. Rovida and T. Hartung, "Re-Evaluation of Animal Numbers and Costs for in Vivo Tests to Accomplish REACH Legislation Requirements for Chemicals: A Report by the Transatlantic Think Tank for Toxicology (T(4))," *ALTEX* 26, no. 3 (2009): 187–208; T. Hartung and C. Rovida, "Chemical Regulators Have Overreached," *Nature* 460 (2009): 1080–81.

78. Schapiro, *Exposed*, 23–35.

79. Reps. John Dingell and Bart Stupak, Comm. on Energy and Commerce, to Hon. Andrew von Eschenbach, commissioner, FDA, January 17, 2008.

80. Ibid.

81. FDA, "Draft Assessment of Bisphenol A for Use in Food Contact Applications," August 14, 2008.

82. R. W. Tyl et al., "Three-Generation Reproductive Toxicity Study of Dietary Bisphenol A in Cd Sprague-Dawley Rats," *Toxicological Sciences* 68 (2002): 121–46; R. W. Tyl et al., "Two-Generation Reproductive Toxicity

Study of Dietary Bisphenol A in Cd-1 (Swiss) Mice," *Toxicological Sciences* 104 (2008): 362–84.

83. FDA, "Draft Assessment."

84. Division of Food Contact Notification, Michelle L. Twarkoski, to Food Master File (FMF) 580-Administrative Record, memorandum, May 27, 2008, www.fda.gov/ohrms/dockets/ac/08/briefing/2008–0038b1_01_10_FDA%20 Reference%20Material-FDA%20Memo%20Develop.pdf. David Michaels details the work of Exponent in his book *Doubt Is Their Product: How Industry's Assault on Science Threatens Your Health* (Oxford: Oxford University Press, 2008), writing, "While some may exist, I have yet to see an Exponent study that does not support the conclusion needed by the corporation or trade association that is paying the bill" (47).

85. Rust and Kissinger, "FDA Relied Heavily on BPA Lobby."

86. The consistency of the plastics trade association's relationship with the FDA was noted by Jerome Heckman, who served as lead counsel to the Society of the Plastics Industry for over fifty years. Jerome Heckman, Heckman and Keller, LLC, interview by author, July 15, 2010.

87. Michelle Twarkoski, formerly with the FDA working on the BPA assessment, joined Keller and Heckman in 2012; www.khlaw.com/ Michelle-Twaroski.

88. John Rost, chair, North American Metal Packaging Alliance, presentation at American Enterprise Institute Center for Regulatory Studies, "The Science and Policy of BPA," June 9, 2010; author attended.

89. Anna Beronius et al., "Risk to All or None? A Comparative Analysis of Controversies in the Health Risk Assessment of Bisphenol A," *Reproductive Toxicology* 29 (2010): 132–46.

90. European Food Safety Authority [EFSA], "Opinion of the Scientific Panel on Food Additives, Flavouring, Processing Aids and Materials in Contact with Food (Afc) Related to 2,2-Bis(4-Hydroxyphenyl)Propane," *EFSA Journal* 428 (2006): 1–75.

91. Meeting of the Bisphenol A Subcommittee of the Science Board to the FDA, September 16, 2008, Rockville, MD; author attended; S.A. Vogel, "The Politics of Plastics: The Making and Unmaking of Bisphenol A 'Safety,'" *American Journal of Public Health* 99, suppl. 3 (2009): S559–66.

92. Office of the Press Secretary, White House, Barack Obama, memorandum for the heads of executive departments and agencies, subject: Scientific Integrity, March 9, 2009.

93. Bob Grant, "More Regulatory Science: FDA Chief," TheScientist. com, September 17, 2009, www.the-scientist.com/blog/display/55984; Margaret Hamburg, FDA commissioner, keynote address, Endocrine Society Annual Meeting, June 12, 2009, Washington, DC, www.fda.gov/NewsEvents/ Speeches/ucm169551.htm.

94. Presentation by Daniel Doerge, National Center for Toxicological Research, Meeting of the Science Board of the FDA, August 17, 2009; author attended.

95. Larger numbers of animals would allow for control of "litter effects," differences that can occur across litters of genetically identical animals. Such differences can have significant effects, particularly on behavioral endpoints.

96. NIH, "NIEHS Awards Recovery Act Funds to Address Bisphenol A Research Gaps," press release, October 28, 2009.

97. FDA, Public Health Focus, "Bisphenol A," January 2010, www.fda.gov/NewsEvents/PublicHealthFocus/ucm064437.htm.

98. EPA, "Bisphenol A Action Plan," issued March 29, 2010, www.epa.gov/oppt/existingchemicals/pubs/actionplans/bpa.html.

99. Administrator Lisa Jackson, Columbia University Center for Children's Environmental Health, "Translating Science to Policy," March 30, 2009; author attended.

100. The EPA argues that the "chemical action plans" fall under the agency's rule-making authority under section 5(b)(4) of TSCA. EPA, "Bisphenol A Action Plan."

101. Ibid.

102. C. Liao and K. Kannan, "Widespread Occurrence of Bisphenol A in Paper and Paper Products: Implications for Human Exposure," *Environmental Science and Technology* 45 (2011): 9372–79; EPA, Design for the Environment, "BPA Alternatives in Thermal Paper Partnership," n.d., www.epa.gov/dfe/pubs/projects/bpa/index.htm (accessed October 12, 2011).

103. Lindsay Layton, "Advocates Run Ad Urging Senate to Pass Food Safety Bill," *Washington Post*, July 11, 2010, A2.

104. Lindsay Layton, "Delay of Food Safety Bill Stirs Tensions between House and Senate Democrats," *Washington Post*, July 22, 2010, www.washingtonpost.com/wp-dyn/content/article/2010/07/20/AR2010072004163.html?nav=emailpage.

105. Henrik Hoegh quoted in Rory Harrington, "Denmark Bans Bisphenol A in Food Packaging for Young Children," *Food Production Daily*, March 30, 2010, www.foodproductiondaily.com/Quality-Safety/Denmark-bans-bisphenol-A-in-food-packaging-for-young-children.

106. Helena Bottemiller, "France Bans BPA in Baby Bottles," *Food Safety News*, May 19, 2010, www.foodsafetynews.com/2010/05/france-bans-bpa-in-baby-bottles/; Mark Astley, "EFSA Begins BPA Report Review," FoodProductionDaily.com, October 20, 2011, www.foodproductiondaily.com/Quality-Safety/EFSA-begins-BPA-report-review.

107. Rory Harrington, "EC Not Ruling Out Action against BPA, Plans to Curb Infant Exposure," FoodProductionDaily.com, October 8, 2010, www.foodproductiondaily.com/Quality-Safety/EC-not-ruling-out-action-against-BPA-plans-to-curb-infant-exposure; EFSA, "Scientific Opinion on Bisphenol A: Evaluation of a Study Investigating Its Neurodevelopmental Toxicity, Review of Recent Scientific Literature on Its Toxicity and Advice on the Danish Risk Assessment of Bisphenol A," *EFSA Journal* 8, no. 9 (2010): 1829.

108. EFSA, "Scientific Opinion."

109. *Toxicological and Health Aspects of Bisphenol A, Report of Joint FAO/WHO Expert Meeting, November 2–5, 2010, and Report of Stakeholder Meeting on Bisphenol A, November 1, 2010, Ottawa, Canada* (Geneva: World Health Organization, 2011), www.who.int/foodsafety/chem/chemicals/bisphenol/en/index.html (accessed September 6, 2011), 28.

110. Ibid.

111. "American Chemistry Council; Filing of Food Additive Petition," *Federal Register*, doc. 2012–3744, February 16, 2012, www.federalregister.gov/articles/2012/02/17/2012-3744/american-chemistry-council-filing-of-food-additive-petition#h-4 (accessed February 27, 2012).

112. C. Liao et al., "Bisphenol S, a New Bisphenol Analogue, in Paper Products and Currency Bills and Its Association with Bisphenol A Residues," *Environmental Science and Technology*, May 16, 2012, doi: 10.1021/es300876n.

113. Daniel M. Fox, *The Convergence of Science and Governance: Research, Health Policy, and American States* (Berkeley: University of California Press, 2010), 110.

114. Julian P. T. Higgins and Sally Green, eds., *Cochrane Handbook for Systematic Reviews of Intervention, Version 5.1.0,* updated March 2011, Cochrane Collaboration, www.cochrane-handbook.org/.

115. Drummond Rennie and Iain Chalmers, "Assessing Authority," *Journal of the American Medical Association* 301 (2009): 1819–21; Cochrane Collaboration, "The Collaboration's Funders," n.d., www.cochrane.org/about-us/funding-support (accessed August 5, 2010).

116. Rennie and Chalmers, "Assessing Authority"; Cochrane Collaboration, "Collaboration's Funders."

117. Stefan Timmermans and Aaron Mauck, "The Promises and Pitfalls of Evidence-Based Medicine," *Health Affairs* 24 (2005): 18–28.

118. Lisa Bero and Drummond Rennie, "The Cochrane Collaboration: Preparing, Maintaining, and Disseminating Systematic Reviews of the Effects of Health Care," *Journal of the American Medical Association* 274 (1995): 1935–38.

119. G. H. Guyatt et al., "GRADE: An Emerging Consensus on Rating Quality of Evidence and Strength of Recommendations," *British Medical Journal* 336 (2008): 924–26; G. H. Guyatt et al., "What Is 'Quality of Evidence' and Why Is It Important to Clinicians?" *British Medical Journal* 336 (2008): 995–98.

120. Tracey J. Woodruff, Patrice Sutton, and the Navigation Guide Work Group, "An Evidence-Based Medicine Methodology to Bridge the Gap between Clinical and Environmental Health Sciences," *Health Affairs* 30 (May 30, 2011): 931–37.

121. While I was working for the Johnson Family Foundation, it provided financial support for a meeting of the Navigation Guide Science Working Group, established to further develop the methodology and implementation of the Navigation Guide.

122. Higgins and Green, *Cochrane Handbook*, sec. 8.2.1.

123. Health Effects Institute, "What Is the Health Effects Institute?," 2010, www.healtheffects.org/about.htm.

EPILOGUE

1. *The Simpsons,* "Pranks and Greens," season 21, episode 6, aired November 22, 2009, FOX Television.

2. Ulrich Beck, *World at Risk* (Cambridge: Polity Press, 2007), 27.

3. National Research Council [NRC], Committee on Improving Risk Analysis Approaches Used by the U.S. EPA, *Science and Decisions: Advancing Risk Assessment,* Board on Environmental Studies and Toxicology, Division on Earth and Life Sciences (Washington, DC: National Academy Press, 2009).

4. Garry D. Brewer, "Five 'Easy' Questions," *Science* 325 (2009): 1075–76.

5. NRC, *Science and Decisions,* 9.

6. Ibid., 9–10.

7. Maria Hegstad, "Activists Urge EPA to Halt Scientists' Participation in Industry Consortia," *Inside EPA,* December 15, 2010.

8. NRC, Committee on Toxicity Testing and Assessment of Environmental Agents, *Toxicity Testing in the 21st Century: A Vision and a Strategy,* Board on Environmental Studies and Toxicology, Institute for Laboratory Animal Research, Division on Earth and Life Sciences (Washington, DC: National Academy Press, 2007).

9. L. S. Birnbaum and W. S. Stokes, "Safety Testing: Moving toward Alternative Methods," *Environmental Health Perspectives* 118 (2010): A12–13; L. R. Rhomberg, "Toxicity Testing in the 21st Century: How Will It Affect Risk Assessment?," *Journal of Toxicology and Environmental Health B: Critical Reviews* 13 (2010): 361–75.

10. R. Judson et al., "The Toxicity Data Landscape for Environmental Chemicals," *Environmental Health Perspectives* 117 (2009): 685–95.

11. Clean Production Action, "Green Screen for Safer Chemicals," white paper, August 2010, http://www.cleanproduction.org/library/cpa-fact%20grscreen_Jan09_final.pdf.

12. Speech by John Warner, "Green Chemistry and Environmental Health: Problems Meet Solutions," November 10–12, 2008, hosted by Advancing Green Chemistry, Environmental Health Sciences, and University of California, Irvine; author attended.

13. Jim Hutchinson, comment at Seattle, WA, meeting hosted by the Bullitt Foundation, March 8, 2010; author attended.

Index

Page references in italics indicate figures.

Abbott Laboratories, 49, 229n13
Abernathy, Thomas, *19*, 222–23n24
ACC. *See* American Chemistry
  Council (ACC; earlier,
  Manufacturing Chemists'
  Association [MCA] and Chemical
  Manufacturers Association)
acrylonitrile: efforts to set standards
  for, 70; EPA restriction on, 128;
  lowered tolerance limits for, 72;
  questions about safety, 60, 67–68,
  82
Ad Hoc Committee on the Evaluation
  of Low Levels of Environmental
  Chemical Carcinogens, 55–56
Administrative Procedures Act (APA,
  1946), 35, 227n82
The Advancement of Sound Science
  Coalition (TASSC), 127–28, 129,
  160, 249n75
AFL-CIO, 73, 119–20
Agent Orange, 128, 235n119
Aggregated Computational
  Toxicology Resource, 216
agriculture: crop yield and food
  price issues in, 18; atrazine issues,
  168, 261n121. *See also* American
  Crop Protection Association; food
  manufacturers and production;
  livestock; pesticides

AIHA (American Industrial Hygiene
  Association), 82
AIHC (American Industrial Health
  Council), 72, 73, 74
air pollution studies, 168
air quality standards, 64, 232n71
Alar episode, 124, 248n63
Albert, Roy, 70
Alcoa Corporation, 194
aldrin, 18
Allen, Edgar, 27
Altria Corporation, 127–28. *See also*
  Philip Morris company
American Bar Association, 36,
  227n82
American Cancer Society, 98
American Chemistry Council (ACC;
  earlier, Manufacturing Chemists'
  Association [MCA] and Chemical
  Manufacturers Association):
  amicus brief filed in *Daubert* case,
  127; anti-BPA demands and, 194;
  BPA review supported by, 202;
  CERHR review panel and, 187,
  189; chemicals in food legislative
  committee and, 23; Chemicals in
  Foods Committee of, 38; cost-
  benefit analysis and risk assessment
  promoted by, 64, 70; Data Quality
  Act utilized by, 168; Delaney clause